RURAL DEMOCRACY

RURAL DEMOCRACY

Family Farmers and Politics

in Western Washington,

1890–1925

᠊᠊ᢒᡱᡈᡪ᠊᠊

MARILYN P. WATKINS

Cornell University Press

ITHACA AND LONDON

First published 1995 by Cornell University Press.

Printed in the United States of America

⊗ The paper in this book meets the minimum requirements of the American National Standard for Information Sciences— Permanence of Paper for Printed Library Materials, ANSI Z39.48-1984.

Library of Congress Cataloging-in-Publication Data

Watkins, Marilyn P. (Marilyn Patricia), 1956–
Rural democracy : family farmers and politics in western Washington, 1890–1925 / Marilyn P. Watkins.
    p. cm.
Originally presented as the author's thesis (Ph. D.—University of Michigan).
Includes bibliographical references and index.
ISBN 0-8014-3073-9 (cloth : alk. paper)
    1. Lewis County (Wash.)—Politics and government. 2. Agriculture and politics—Washington (State)—Lewis County—History—19th century. 3. Agriculture and politics—Washington (State)—Lewis County—History—20th century. 4. Family farms—Washington (State)—Lewis County—History—19th century. 5. Family farms—Washington (State)—Lewis County—History—20th century. I. Title.
F897.L6W38 1995
306.2'09797'8209041—dc20   95-31253

F
897
1 L6
W 38
1995

Contents

Illustrations

Tables

Preface

I was initially drawn to the study of rural politics by my interest in women's history. During the past decade a great deal of fine scholarship has been published on farm women, sparked by the pioneering work of historians such as Joan Jensen and Susan Armitage. But when I started graduate school in the early 1980s, white, urban middle-class women received much of the attention of women's historians. I wondered whether rural women—who still outnumbered town women in the late nineteenth century—were able to take part in politics without access to the variety of female associations through which their urban sisters entered civic affairs. When I began to explore rural politics, I was struck by the lack of attention scholars had given to women and issues of gender. The Grange and the Farmers' Alliance had taken the highly unusual position for the nineteenth century of insisting on the equal inclusion of women. The Populists and other farmer-based movements had gone on to take progressive stances on women's rights. Understanding gender roles within these movements thus seemed crucial to understanding the involvement of rural people in the political process. I also became increasingly interested in whether and how twentieth-century farmer-based movements were related to those of the nineteenth century. This book was originally conceived to address these two issues.

I owe much to the rigorous training, intellectual excitement, generous encouragement, and financial support of the Department of History at the University of Michigan, its faculty, and students, and take this opportunity to offer my thanks. I am particularly grateful to Terrence McDonald, for his steadfast confidence in me and for often showing me where I was going before I knew myself. Maris Vinovskis, Carol Karlsen, and Howard Kimeldorf offered perceptive criticism and enthusiastic support. Robert Berkhofer and Sidney Fine were helpful in the early stages.

Nancy Grey Osterud, Sarah Deutsch, Richard White, John Findlay, Karen Blair, Sue Armitage, and Kathleen Underwood read earlier versions of the manuscript, and I am grateful to them all for their encouragement and insightful comments. Peter Agree of Cornell University Press has been kind, patient, and supportive in guiding me through the publication process.

A community study of this sort would not be possible without the assistance of the people still living in that community. Members and officers of the Granges, fraternal lodges, and churches of south Lewis county located old boxes full of records, gave me work space— sometimes in their own kitchens—to go through them, answered my questions, and often fed me as well. Volunteers and staff of the Lewis County Historical Museum, the Washington State Historical Society, the Washington Archives, the Washington State Grange, the Pacific Northwest Regional Archives of the National Archives, the Pacific Northwest Methodist Archives, and the Seattle Public Library were also extremely helpful. The University of Washington Libraries provide generous assistance to visiting scholars, and I am especially grateful to the staff of Special Collections, Manuscript Collections, Newspapers and Microfilms, and the Inter-Library Borrowing Service.

I have been blessed with a large and supportive extended family and circle of friends who have been there when needed through the years. I especially thank my parents, Keith and Billie Watkins, for my earliest academic training and their cheerful acceptance of the various paths their children have chosen. My father-in-law, Howard Ulberg, lent me his laptop computer. My friend and neighbor Mary Marrah helped finish the maps in this book when I was about ready to give up on them. Cy Ulberg provided most of the financial support and much of the computer expertise that made this book possible, in addition to being a wonderful partner. Carl and Erik made sure I kept this work in perspective and have taught me many things no one else could have.

MARILYN P. WATKINS

Seattle, Washington

RURAL DEMOCRACY

⟨ 1 ⟩

Introduction: Agrarian Activism,

Gender, and Lewis County

In the spring of 1918, three hundred farmers gathered angrily for the quarterly meeting of the Lewis County Pomona Grange. Less than two weeks earlier, townsmen from the trading centers of Winlock and Toledo had tarred and feathered two members of the Nonpartisan League who were organizing in farming districts. Lewis County Grange members unanimously condemned the persecution of League organizers as an attack on democracy and constitutional rights. Many townspeople, on the other hand, proclaimed the vigilantes heroic patriots who had rid the area of dangerous agitators, since the Nonpartisan League was widely suspected of disloyalty in a time of war. Local newspapers supported the vigilantes, and the sheriff made no move to investigate the attacks.

The conflict between farmers and townspeople in Lewis county over the meaning of democracy and its political expression was of long standing. It had been going on at least since the 1890s, when the Populist party disrupted two-party politics and demanded a reordering of the national economy. Despite such old antagonisms, in February 1923 the people of Winlock and the surrounding rural areas rallied together in support of a new state road, with Grange members, business people, and mill workers eating, singing, and cheering side by side. In doing so they exhibited another facet of their common history: the solidarity and commitment to their larger community that had bound them together through all the years of political turmoil.

In this book I focus on the political culture that developed in a portion of south-central Lewis county during several waves of agrarian activism between 1890 and 1925. Most of the area joined the

1

Populist crusade for a cooperative commonwealth during the 1890s. The primarily farming precincts continued supporting third parties that called for democratic reforms and economic restructuring into the twentieth century, giving a high percentage of votes to the Socialist, Farmer-Labor, and Progressive parties between 1908 and 1924. The small towns, by contrast, remained largely Republican after 1900. Farmers of the area received most of their political education in the Farmers' Alliance in the 1890s and in the Grange after the turn of the century. These same organizations were also at the center of social life in most rural neighborhoods. The community basis of farmers' politics gave rural men and women the strength to hold firm to their demands for reform, even as the political spectrum narrowed and attacks on suspected radicals became increasingly common in the late 1910s. The community base also provided rural women access to partisan politics at a time when it was generally considered a male domain. Within the Alliance and Grange, political discussion and campaigning came to be seen as responsibilities shared by men and women. At the same time, women and men held different and complementary roles in the social and economic components of organizational life, making the contributions of both sexes vital to the long-term health of the reform movement.

Historians usually deal separately with the agrarian activism of the 1890s and 1910s, drawing at best superficial ties between movements. Indeed, national party ideology and leadership as well as the geographic centers of strength of these movements did differ. The United States was changing in profound ways during this period, with dramatic innovations in industry, an influx of immigrants from diverse cultures, the tremendous growth of cities, and the steady increase in scope and power of the government. Agriculture, too, was being transformed with the expansion of domestic and international markets, and the spreading of new technology and agricultural science which greatly increased production, at the same time threatening the existence of the family farm. Lewis county farmers participated in these changes in the early twentieth century, benefiting from the government-financed Extension Service, increasingly specializing in poultry and dairy production for the urban market, and establishing successful marketing and processing cooperatives that allowed them to maintain a competitive position. Although their specific political proposals shifted along with economic and political changes, in many ways the goals of Lewis county farmers remained the same: that government, representing the interests of all the people, should control powerful trusts and corporations, and that farmers, along with all producers, should receive a just return on their labor and exercise control over their own affairs, within the bounds of community responsibility and an industrial, democratic society.

Lewis county farmers proudly affirmed their American heritage and their commitment to the Constitution and the values expressed in the Declaration of Independence. Their platform posed a radical challenge to the political status quo, however, in its ideal of power resting in the hands of a fully mobilized and educated citizenry, and in its insistence that economic justice be included as a goal of government policy along with protection of private property and individual profit. To many townspeople the farmers' attacks on the injustices of monopoly capitalism sounded socialistic, un-American, and with the country's entrance into World War I, downright treasonous. At the same time, neighborly ties, the desire for community improvement, and common cultural life overcame many of the divisions between farmers and small-town residents, and encouraged even the most radical farmers to find ways to compromise with the prevailing culture. Focusing on one locality from the founding of the Populist party through the Farmer-Labor movement allows us to see these continuing ties of community as well as the unity of this period of agrarian protest.

Farmers developed economic cooperatives within this tension between the community culture they shared with townspeople and their politicized consciousness of themselves as producers. The cooperative movement that blossomed in the twentieth century grew from the same desire for economic independence and a just return on labor that inspired third-party politics, and was encouraged by the Grange. Town business leaders and civic organizations, the Farm Bureau, and government agents also actively promoted the cooperatives and at times took credit for their success. Participation in cooperatives helped farmers reach the economic prosperity that had long been their goal, but did not cause them to reject progressive politics and alliances with organized labor as many outside promoters had hoped. Rather, farmers chose which aspects of government and business assistance they were willing to accept and insisted on maintaining control in their own hands, despite pressures to yield to the leadership of "experts." The extremely successful experience with cooperatives in Lewis county also failed to lead to the more radicalized consciousness that some historians believe had resulted from the collapse of the Farmers' Alliance's economic initiatives in the South in the 1880s and 1890s.[1] The fact that the cooperatives in Lewis county were organizationally separate from the Grange, the center of community social and political life, allowed farmers to enter into economic alliances with their political opponents, without giving up their base of independence.

The redefinitions of gender roles that occurred in the Alliance and Grange also had ambiguous consequences. Members of the farm organizations came to see politics as a responsibility shared by men and women in a world otherwise highly organized by gender. Rural women gained a political voice and a level of power in the Alliance and Grange they were denied in other community associations and in the traditional political parties even after women won the right to vote. However, the continued sexual division of labor in economic life and in other social and cultural associations greatly lessened the impact of that achievement.

Although my focus here is on one small group of communities, farm families in western Washington state participated in a much broader movement for political and economic change. During the late nineteenth and early twentieth centuries, the United States became one of the world's major industrial powers. The federal government encouraged industrial growth through such policies as land grants to railroad companies, tariffs on imported manufactured goods to protect emerging domestic industries, and the suppression of labor unrest. The rapid economic and social changes brought political changes as well, but only after much conflict and struggle. From the 1870s to the 1930s, farmers outside the northeastern economic core of the nation organized a series of political movements that challenged the priorities of the emerging system. The Greenback, Populist, Socialist, Nonpartisan, and Farmer-Labor movements differed in their centers of strength and specific proposals, but all shared the goal of securing government protection from the vagaries of an international capitalist economy and the abuses of industrial corporations. During the same period, industrial workers, urban women, African Americans, and social reformers also fought to secure a more equitable division of wealth and power in the new industrial economy. These "struggles for justice," as Alan Dawley terms them, pushed the government away from its long-held laissez-faire and probusiness stance to a much more interventionist position by the 1930s.[2]

Looking from a historical distance at the nation as a whole, we may see this variety of movements as more continuous and coherent than it seemed at the time. In any given place or group of people, participation was often fragmented. That Lewis county farmers took part in the agrarian wing of this struggle for over thirty years is due in part to the place of the West in the national economy. The various waves of agrarian activism between 1870 and 1930 had different centers of strength, but all were concentrated in the nation's periphery—the South, the Plains states, and the far West—outside the heavily industrialized northeastern and Great Lakes core. Farmers near the big industrial cities had a ready domestic market for their produce and believed their financial and political interests were tied to the success of business and industrial growth. Thus although political scientists and historians have found evidence of tensions between town and countryside throughout the nation, farmers of the Northeast and as far west as Illinois tended to support the Republican party, the party of prosperity and business growth, from the end of the Civil War to the New Deal and indeed beyond. Farmers outside the core region, in contrast, raised commodities for both the national and international markets. Like colonies serving the mother country, the states of the periphery relied heavily on investments and manufactured goods from the core, while furnishing raw materials of fiber, grain, and minerals that added more to the power and wealth of faraway capitalists than to themselves. The railroad and the eastern banker were the most visible and despised symbols to farmers of their dependent status.[3]

To Westerners particularly, the federal government was an obvious tool to correct this regional imbalance of power. The federal army had conquered the land and banished the native inhabitants to isolated reservations; the government had provided enormous land grants to promote construction of the railroads that, for all their highhanded policies, had opened the land to white settlement and provided access to markets; the government was supposed to be of the people and for the people and was the only entity with the scope and power to rein in the mighty trusts.

Although the Populist party rallied strong support in both the South and West in the 1890s, by the early twentieth century, as Richard M. Valelly argues, third-party radicalism had the chance to succeed only in the West. Industrial workers and farmers throughout the country sought federal protection, but their demands were largely ignored by a government that supported capitalist industrial expansion. The two major parties could also afford to ignore reform demands, since the Democrats had secure hold of the South and the Republicans comfortably dominated the North. In western states, however, traditional party organizations were relatively weak. There radical political entrepreneurs could tap reform sentiment and break into the political arena through a third party. If they gained control of state governments, these movements could enact meaningful reforms. North Dakota's Nonpartisan League and Minnesota's Farmer-Labor party were the most successful examples of such state-level radicalism. With the New Deal, however, the federal government itself took on reform with a bureaucracy capable of reaching into the lives of individual Americans, thus preempting the political space formerly occupied by radical third parties.[4] The same possibility for third-party success also existed in the South in the late nineteenth century. While the white landed aristocracy was still working to reestablish political control following the devastation of the Civil War, the Populist party could gain a foothold, but with the consolidation of elite power in the Democratic party in the 1890s, that possibility rapidly diminished.[5]

Economic grievances have been central to historians' understandings of the Populist movement. In 1931 John D. Hicks laid out what is perhaps still the most widely accepted view of the Populists in *The Populist Revolt*. Hicks viewed the agrarian revolt as a consequence of the rapid closing of the frontier and "the struggle to save agricultural America from the devouring jaws of industrial America."[6] During the 1870s and 1880s, farmers had brought tens of thousands of new acres under cultivation as they expanded onto the Great Plains, frequently contracting debts in doing so. Southern farmers, in the meantime, struggled to overcome the destruction of the Civil War. Many yeoman farmers of the South lost their economic independence as they fell under the crop-lien system, pledging the next year's cotton crop for necessary supplies. Cotton and wheat farmers of the South and Midwest were locked into debts they could not repay as crop prices fell with increased production and general de-

flation, yet they were forced to pay higher rates to railroad companies and other monopolies. With the organization of the Farmers'
Alliance, farmers found their political voice. Drought in the late
1880s and the deep and prolonged depression of the 1890s gave
impetus to their movement and added to their base of support. The
movement died as the depression ended and new gold discoveries
led to an expanded money supply and higher agricultural prices.
But Hicks concluded that many of the Populists' specific reforms
were ultimately accepted by the Progressive reformers of the twentieth century, along with much of their overall philosophy: that government must restrain the greed of those who profit from the poor,
and that the people—not special interests—must control the government.[7]

For much of the past forty years, the study of Populism has been
dominated by attempts to refute or support the latest reinterpreta-
tion of the movement. In the 1950s, searching for an explanation
of McCarthy's ability to disrupt the orderly working of the system,
Richard Hofstadter published a sweeping condemnation of the Populists as carriers of the ugly streak of bigotry and intolerance in
American politics. He found the motivation for the movement in
status anxiety rather than economic misery. Like Hicks, Hofstadter
saw the Populists—along with the Progressives of the following decades—in an inevitably losing battle with industrialization, but he
interpreted the movements as irrationally looking backward to a
mythical past rather than coping with reality, and as responding to
their followers' anxieties with hate politics.[8] For the next twenty
years, scholars refuted Hofstadter's thesis, arguing that the Populists
(and Progressives) had practical proposals to deal with real economic problems and political injustices and had exhibited no more
prejudice and intolerance than the mainstream of American society.
Others pointed out that the political bases of support for reform at
the turn of the century and for McCarthy in the 1950s were quite
different.[9]

The debate was diverted in different directions with the 1976 publication of Lawrence Goodwyn's passionate tribute to the Populists
as America's last great (though failed) chance for true democracy.
Goodwyn described the unique movement culture that developed
in the Farmers' Alliance as farmers met regularly in an atmosphere
that encouraged participation and debate, learned hard lessons
about economic structure and power through their attempts to es-

tablish buying and selling cooperatives, and spread their message through traveling lecturers. In the Alliance, farmers developed a vision of a democratic and just America that was the source of strength of the People's party. To Goodwyn, it was their overall vision and not specific reforms that comprised the essence of Populism, and thus, unlike Hicks, he sees it as (regrettably) an ultimate failure.[10]

Many have disputed the specific chronology of Goodwyn's account, and particularly the importance he places on experiences with economic cooperatives in developing political consciousness and goals, and his dismissal of most midwestern Populists who did not participate in the great cooperative experiments as members of a "shadow movement."[11] Goodwyn's emphasis on culture and its development in the Farmers' Alliance has, however, been widely acknowledged as important. Robert McMath has argued that the traditions of community cooperation, responsibility, and independence that infused Populism were well established among southern and midwestern farmers before the organization of the Alliance. The Alliance was then able to build on these preexisting republican traditions in creating its movement culture.[12] Sociologists and political scientists have also been attracted to Populism, studying organization, leadership, and class standing to determine how resources were mobilized and why popular support ultimately fizzled.[13]

Given the long-standing scholarly interest in the movements, we know quite a bit about different groups of agrarian radicals active at different times in different parts of the country. It is clear that no single factor explains all of these movements, but rather a combination of economic, social, and political factors. It is also clear that the clustering of agrarian unrest between the end of the Civil War and the Depression of the 1930s is related to the disadvantaged economic position of small-holding farmers as national policies promoted rapid industrialization, and to the structure of the political system that both closed off other routes of protest and allowed third-party challenges a chance of success, however limited, at the state and regional levels.[14]

What can this study of yet another group of agrarian rebels in an admittedly obscure part of the country tell us that we do not already know? Lewis county farmers were neither typical of agrarian radicals nor so exceptional as to require special explanation. They did, however, show a remarkable loyalty to a third-party critique of industrial capitalism as it existed in the decades around the turn of the century. Most studies of agrarian activism have focused on one movement and the actions of a specific group of leaders, stressing the distinctions between that movement and what came before or after.[15] Often we fail, therefore, to see the common links. The story of Lewis county farmers allows us to see more clearly than more fragmented examples the continuing role that rural culture played in the spawning and nurture of political protest. Their activism was not the result of any single crisis in the regional or national economy. These were reasonably prosperous people, by and large free from debt, already integrated into the international market, and able to respond with some flexibility to new opportunities. They shared with small-scale farmers in many parts of the country a culture that valued independence and neighborly reliance, private ownership of property and community responsibility, physical labor and prosperity. Unlike farming regions throughout the South, however, Lewis county rural communities were not divided by sharp class and racial differences that undercut their ability to work together. When the Alliance collapsed along with the Populist party in the late 1890s, the Washington State Grange quickly reorganized farmers. And while the Grange in many parts of the country steered clear of reform politics after the turn of the century, Washington state leaders established strong ties with organized labor and aggressively pursued many of the reforms the People's party had championed. Skillful state leaders, however, were not alone responsible for maintaining the political activism of Lewis county farmers. Within their organizations, local people built a culture of participatory democracy that gave them the courage to challenge existing economic and political structures and try to change them in accordance with their values, even in the face of strong local opposition.

Political culture and its grass-roots manifestation brings us to an area that has been almost entirely neglected by historians of agrarian protest: women's activism and the challenge to prevailing notions of gender in agrarian movements. Despite the scores of volumes written about Populism and to a lesser extent other rural political movements, few scholars give more than passing attention to women's participation or to any issues of gender. This is true even in quite recent works that were researched and written after women's history became a well-established field, and despite numerous indicators pointing to gender as a fruitful issue of study, including: the debate over women's suffrage by Populists; the national recognition of women speakers, writers, and theoreticians such as Mary Elizabeth Lease and Sarah Emery who rose to prominence in the Midwest; the conscious inclusion of women in activist rural organizations like the Grange and Farmers' Alliance; and the women's suffrage campaign and widespread activism among urban women that was going on simultaneously with the period of intense farmer unrest.

Women's history has expanded our very definition of politics by bringing to light the public activities of urban women that often took place outside the world of party conventions and election campaigns. Throughout the nineteenth century, law, custom, and ideology joined to create a powerful barrier to women's participation in "male" party politics. The ideology of domesticity or separate spheres sharply distinguished "natural" female and male qualities. It glorified motherhood and ascribed virtue and the keeping of morality to women within the private realm, while viewing men as naturally competitive and aggressive, suited for leadership and the pursuit of capitalism in the public realm. Although this ideology confined women to a far from equal sphere of activity, women used that very separation as a source of strength, creating a lively and supportive women's culture and gradually pushing their role as guardians of morality into full-scale public activism. Middle-class women organized opposition to drinking, prostitution, and child labor, and both provided and lobbied for a wide range of social services for impoverished women and children. By the time women achieved suffrage nationally in 1920, the meaning and nature of politics had been transformed—in part through women's activism —and the domestic sphere had expanded to the point that the two realms were no longer easily distinguishable.[16]

While the ideology of separate spheres has been accorded a central role in much of the literature on the organized activities of white middle-class urban women, we also have evidence of a great deal of diversity among women's groups and areas where "separate spheres" was not the prevailing paradigm. African American and white working-class women often focused on race or class, rather than gender issues.[17] Recent studies of rural women and men, too, have revealed more mutuality than separation between the sexes in turn-of-the-century America. In their studies of family and work life, Nancy Grey Osterud, Mary Neth, and Karen V. Hansen have all found routine blurring of a sexual division of labor on family farms.

Not only did women help out with "men's work," as has been well documented on the frontier and in wartime, but men helped women with their work, too. Gender-based distinctions and hierarchy did exist, and women shared much in common and relied upon each other, but at the same time, work and social life were frequently mixed-sex. The work of these and other scholars moves beyond the frequent assumption that farmers were men and the farm family was a single unit whose interests were represented by the male farmer, by breaking down the household into individual members and examining the relations—and conflicts—among them.[18]

At this point, we still know very little about women's participation in rural political life. Even though Goodwyn saw the Farmers' Alliance and the movement culture that grew out of it as the heart of Populism, he dealt little with the women involved in that transformative process. Scott G. McNall noted that women constituted about one-quarter of Alliance members in Kansas, but considered them largely irrelevant to political outcomes, aside from the few female speakers and theorists, because they were not voters.[19] While a few historians have looked seriously at women in the Alliance, notably Robert McMath, Julie Roy Jeffrey, and Maryjo Wagner, remarkably little attention has been paid to this area.[20] Studies of upsurges in agrarian radicalism during the first decades of the twentieth century are more scarce, but the record here is not any better—women are often mentioned in passing, but are rarely treated seriously. Most authors assume, furthermore, that the actions and ideologies of men can be understood without reference to the women who lived and worked alongside them.[21]

In her study of hospital workers in the 1970s, Karen Sacks makes clear the distortion of the historical record that can result from ignoring women's involvement. Sacks found a sexual division of labor within the workers' unionization movement, where men served as spokespersons and women as the "centers and sustainers of . . . networks." Although a different time, place, and culture from the family farmers this book focuses on, Sacks recognized this particular role of women as an important aspect of grass-roots activism.[22] A traditional treatment of the unionization drive that focused only on the largely male public spokespersons would have missed a significant element of the movement—the core of community-building that made the drive a "movement" rather than simply an election campaign and sustained it for over ten years. Similarly, studies of agrarian movements that focus on male leaders and discuss the importance of the culture from which they came, without reference to the women who helped make that distinctive culture possible, are missing a key component of those movements.

In Lewis county, the sexual division of labor prevalent in other community associations was followed in the Alliance and Grange as well, with men providing public leadership and women providing much of the food, entertainment, and relief efforts on behalf of members that fostered a sense of family and group solidarity. Women were thus essential to the long-term health of the neighborhood locals, but much less visible in state and national meetings. At the same time, the ideology of sexual equality and the opportunity to participate in political debates and activities in their neighborhood assemblies opened up new possibilities for rural women. The records show clearly that women were just as concerned with "the issues of the day" as the men. Alliance and Grange women were among the pioneers of the mixed-sex political culture that was slowly emerging in the United States and was confirmed with the passage of the Nineteenth Amendment in 1920.

A dense network of voluntary associations in Lewis county, including fraternal lodges, churches, and less formal organizations, made the spreading of new political ideas that resonated with the people fairly easy. When the Farmers' Alliance and Grange organizers came calling, the forms of organization and community-building were already in place, as were the ties among neighbors and friends to facilitate recruitment into a new association. At the same time, those networks and the commitment to the larger community that many people felt also blunted the radicalism of the agrarian political movements. Most people wanted to avoid antagonizing their neighbors and lodge brothers and sisters. The strong ethic of community inspired people both to activism in defense of their traditions and to compromise, especially in a society where occupation and economic standing were too fluid for class to be sharply divisive. Farmers' Alliance members and Grangers acknowledged themselves as members of the "productive classes" and supported striking industrial workers throughout the West with money and resolutions. They railed against those who grew rich from the labor of others, and at times voiced grievances against the merchants of the nearby towns. At the same time, activist farmers recognized that they shared with small-town businessmen the desires for their community to prosper, their produce to sell profitably, and transportation and communication throughout the region to improve. In certain settings, they championed a transformation of social and economic relations in the United States with heated political rhetoric. On other occasions, they set aside differences to work with their political opponents on matters of common interest. This fluid aspect of political culture, like gender relations, is more easily seen at the local than at the state or national levels.

I chose Lewis county as the site of my work because farmers there participated in the movements I was interested in, enough source material could be located for research, and it was convenient to my Seattle base. Lewis county was fairly representative of rural western Washington of the period in its mix of agriculture and logging and its demographic profile. Western Washington farmers shared much in common culturally and economically with farmers throughout the country, but in a nation so diverse in soil and climatic conditions, crop and labor systems, and access to markets and credit, no farmers can really be considered "typical." What makes this study important for people interested in other regions and other movements is not its representativeness but the issues it raises, particularly those concerning the community base of political activism and women's particular contributions to political culture.

My research led me to a variety of sources. Many of the Granges, fraternal lodges, and churches of the period that remain active today allowed me access to their records, including minutes and membership lists. A scattering of manuscript collections, local histories, and printed memoirs, and several interviews with descendants of those involved, added important insights. Manuscripts of the federal census for 1900 and 1910 combined with county tax records provided a basic demographic and economic profile of the area. Local newspapers were an important source for the basic chronology of events, the activities of organizations that did not leave records, and local issues. During the 1890s, Chehalis, the county seat, housed three weekly newspapers with countywide circulations, the Republican *Bee*, the Democratic *Nugget*, and the Populist *People's Advocate*. All three carried boilerplate copy produced elsewhere summarizing national and international news and several locally produced pages of gossip, crop and weather reports, other local news, and strongly partisan editorials and letters. By the early twentieth century, the *People's Ad-*

vocate had become the *Lewis County Advocate* and adopted a generally moderate independent stance, while the other two papers had merged into the Republican *Bee Nugget*. All three papers have been substantially preserved on microfilm, and I used them extensively. Other towns, including Winlock and Toledo, had their own weekly papers, which were staunchly Republican. Centralia briefly supported a Socialist paper from 1913 to 1915, the *Lewis County Clarion*, along with several other more traditional offerings. Unfortunately, only a scattering of issues of the smaller papers from the time period are available today.

Throughout the book I use the term "farmer" somewhat differently from its conventional use both today and at the turn of the century. "Farmer" has generally been used to refer to the head of a farming household, typically male. Census takers in the study area both in 1900 and 1910, when recording the occupations of a married couple living on a farm, would describe the husband as a farmer and the wife as having no occupation. Men joining the Grange in the early twentieth century described themselves as farmers when signing the roll book, but women more often called themselves housewives or farm wives, although occasionally they, too, signed in as farmers. Nevertheless, I use the word "farmer" to refer to all family farm workers. Women and older children clearly provided much of the labor necessary to keep a family farm going. Furthermore, in this region no permanent separate class of tenants existed to divide the farming population, and seasonal wage labor either was drawn from the family farming economy or was migratory and thus not considered part of the community by permanent residents. Rather than having to repeat a cumbersome litany of everyone included in farm work, therefore, I have adopted this unconventional —though certainly not inappropriate—usage.

The next three chapters focus on Lewis county during the 1890s. Chapter 2 details the economic and social organization of the south-central part of the county. In Chapter 3, I discuss the distinctive political culture that developed within the Farmers' Alliance, and particularly the opportunities it presented for women. Chapter 4 examines partisan politics in Lewis county and the interplay of local and national issues. The scene shifts to the twentieth century in Chapter 5, in which I discuss the entrance of the Grange into community life and its involvement in Progressive reform. In Chapter 6

I look at agricultural specialization and the development of marketing cooperatives, and local farmers' response to the Extension Service in light of their political beliefs. Chapter 7 documents the shattering effects of World War I and the Red Scare, and the efforts to mend community relations in their wake.

ҫ 2 ӡ

Rural Community Life:
Lewis County in the 1890s

In the mid-1880s, Judd and Susan Herren came to Lewis county from North Carolina with their six children, Judd's brother, and a party of friends. They rode the train as far as Portland, then took a steamboat to Toledo. The Herrens were in their mid-forties and financially well enough off to purchase 308 acres of prime farmland on Cowlitz Prairie near an old Hudson's Bay trading post and Catholic mission. The family prospered in its new setting. Susan apparently remained occupied primarily with family chores, but Judd was freed by the labor of his wife and children to become a leader in community life. He served as president of the Lewis County Farmers' Alliance through most of the 1890s (with his son Hugh acting as secretary) and was elected vice president of the State Alliance in 1894; in 1901 he was one of the founders of the Toledo Masons, transferring membership from his North Carolina chapter. In addition to farming and public life, Judd also undertook several business ventures. He invested in a freight company carrying goods by steamboat between Toledo and Portland, and with friends who had also made the journey from North Carolina he formed a logging company and planted a hops yard. Judd's brother Sam, in the meantime, practiced law in Chehalis, was elected to the state legislature, and became active in Populist party politics, before moving to Idaho and running a newspaper in the late 1890s.

Judd, Susan, and their children remained on Cowlitz Prairie, among the area's "leading citizens." By 1910, two unmarried children continued to live at home, the son farming alongside his father and the daughter working as a schoolteacher. Another son, named Sam like his uncle, lived nearby on his own farm with his wife and two small children. The younger Sam also ran a store that housed Grange meetings in the early 1910s and was elected the first master of the reorganized Cowlitz Grange in 1920. Some of Judd and Susan's grandchildren still live and farm on Cowlitz Prairie.[1]

While the Herrens may not have been entirely typical, their family history does represent characteristics common to many of the residents of Lewis county at the turn of the century. They came with the railroad, intending from the beginning to both provide for their families and produce for national and international markets. They prospered through hard work and a particular family division of labor. Their ties to the community were reinforced through extended family, long-time friendships, and neighborhood associations. And they joined with their neighbors to promote their common economic and political interests in a variety of ways.

This book focuses on the south-central portion of Lewis county, from the small town of Napavine south to the county line, and east to Silver Creek and Cinebar.[2] It is an area of thick forests, rolling hills, and river valleys, about halfway between Seattle and Portland. The forests are interspersed with a number of prairies, at least some of which were created, according to early white settlers, by local Indians burning the land to encourage growth of food plants and game animals.[3] Another tier of counties to the west separates Lewis county from the Pacific Ocean, and the Cascade Mountains rise along its eastern edge. It shares the generally mild and rainy climate for which western Washington and Oregon are well known.

When the Herrens first moved to Cowlitz Prairie, Lewis county along with the rest of Washington state was just emerging from the frontier. The completion of the transcontinental railroad during the 1880s spurred tremendous population growth. The newcomers completed the conquest of territory from the native inhabitants and decisively established white American cultural, as well as military and political, dominance. The railroad also carried away the vast yields of Washington's forests and farms to national and international markets. The decade of the 1880s climaxed with the achievement of statehood in 1889. Most people across the state looked forward to continued growth and prosperity.

Sixty years earlier, the area had been the home of several bands of Cowlitz Indians who harvested the abundant salmon for their staple food and moved seasonally to gather and sometimes cultivate roots and berries.[4] In 1832 the British-owned Hudson's Bay Company opened a trading post on Cowlitz Prairie near present-day Toledo. The St. Francis Xavier Mission was founded nearby by Roman Catholic missionaries a few years later. Unfortunately, the whites carried diseases for which the Indians had no natural immunity, and many of the Cowlitz and other coastal Indians died in the resulting epidemics.[5]

In the 1850s, Anglo-American settlers began trickling into the newly created Washington territory, encouraged by the final achievement of a boundary settlement with Britain and the generous offer of free land by the United States government. White settlement, however, remained sparse through the 1870s. The remaining Cowlitz Indians refused to sign a treaty ceding their land because they were not provided a reservation of their own; they continued to live along the Cowlitz River and on Alpha Prairie after the Indian wars of the 1850s, gathering food and trading salmon with the growing numbers of whites. Gradually, they were forced by the increasing white population to settle on farms themselves or scatter onto the reservations of related tribes.[6]

Construction of a railroad in the early 1870s between Kalama on the Columbia River and Tacoma on Puget Sound spurred some development, but the Northern Pacific Railroad Company went bankrupt in 1873, leaving a 1,500–mile gap in the proposed transcontinental route. Population in the Northwest remained sparse until a reorganized Northern Pacific completed the transcontinental line in 1883. Then Lewis county's population increased rapidly, growing from 2,600 to 11,499 between 1880 and 1890. The small settlements of the early 1880s were booming towns by 1890, especially the ones located near the railroad. (See Map 1 for a view of the study area.) Just a few miles north of the county seat of Chehalis was Centralia, the largest town, which grew from 200 people in 1884 to 5,000 by 1892. One traveler visiting in 1890 reported counting 107 buildings under construction from his Centralia hotel window.[7]

Napavine, seven miles south of Chehalis, was divided into town lots by two enterprising men when the railroad first came through in 1872. By the late 1890s the town had five lumber mills, a flour mill, four churches, and several businesses, and served a thriving agricultural region.[8] Winlock (see Figure 1) had been only a post office until the Northern Pacific opened a station there in the early 1870s. By 1890, Winlock boasted a population of 650, saw and plan-

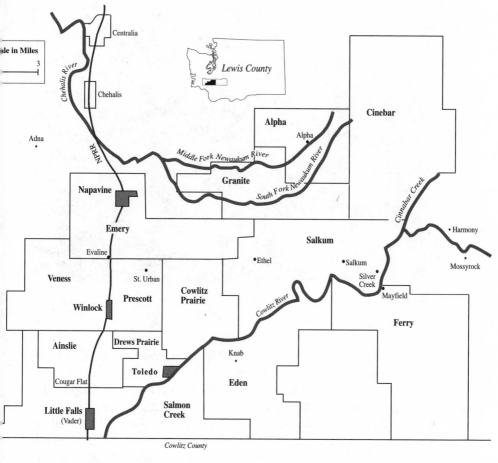

Map 1. Study area precincts, 1900

ing mills, a sash and door factory, a bank, three saloons and over a dozen other business establishments, a newspaper, Baptist and Methodist Episcopal churches, and a grade school with 250 pupils. Winlock's Masonic lodge moved into a "commodious" hall in September 1890, and the Oddfellows built their own hall the next year.[9] Winlock residents were hopeful enough of their future to petition—unsuccessfully—to have the county seat moved from Chehalis. The destruction of the business district by fire in August of 1893 caused only a temporary setback, despite the fact that few owners carried insurance and a severe nationwide depression had settled in. By the

Figure 1. View near Winlock, Washington, early 20th century. Photograph courtesy of Lewis County Historical Museum, Chehalis, Wash.

next February, newspapers reported the town had been rebuilt and no sign of destruction remained.[10]

South of Winlock near the county line was the town of Little Falls, renamed Vader in the twentieth century. It had a clay and brick factory as well as mills, several churches, small businesses, and an Oddfellows lodge.[11] A mill opened between Winlock and Little Falls in 1874 and the town of Ainslie developed around it. In 1892 the Ainslie mill employed 300 men, but through the hard times of the mid-1890s the mill changed hands several times and was frequently in financial difficulty. When the mill finally closed for good around 1897, the town of Ainslie disappeared along with it.[12] Toledo (see Figure 2) was the only town of any size that was not on the rail line. The town began in the late 1870s as the terminus of a steamboat line between the Cowlitz Prairie settlement and Portland. Toledo, too, had shingle and saw mills, but it grew primarily as a farm market town of shops and churches.[13]

Several rural neighborhoods relied on the shipping facilities, stores, churches, and services of these towns. Cowlitz Prairie, which lies just northeast of Toledo on the north side of the Cowlitz River,

Figure 2. Toledo, Washington, street scene, early 20th century. Photograph courtesy of Lewis County Historical Museum, Chehalis, Wash.

was a fertile and thriving agricultural region with gently rolling hills and stunning views of the Cascade mountain peaks. Eden Prairie, later named Layton Prairie after a prominent family, is across the river and directly east of Toledo. Its post office was named Knab after the first postmaster. Between Napavine and Winlock lie the districts known as St. Urban and Evaline. The logged-off lands south of Winlock were purchased early in the twentieth century, mainly by Finnish immigrants who took up poultry raising on what turned out to be marginal crop land.

The area to the east of the railroad was much less densely settled. The neighborhood of Ethel lies just upriver from Cowlitz Prairie, with Alpha Prairie to the north. The villages of Salkum, Silver Creek, Mayfield, and Ferry, further east, provided minimal services. Salkum and Silver Creek were sixteen and nineteen miles respectively east of Napavine, the closest railway station. Mayfield and Ferry were a few miles further east on the Cowlitz River and supported a lumber mill. These two town centers were only one-half mile apart, and the boosters of each competed vigorously, at times bitterly, over which would house the post office and other facilities.[14] Although a striking natural setting, this area was more rugged than the lower Cowlitz valley, with steeper hills and fewer prairies breaking up the forests. Fertile farmland and timber and mining regions lay further to the east, but no major population centers to attract railroad companies, and the Cowlitz River was not navigable beyond Toledo. Most residents relied on mixed farming to earn a living, but without as ready access to the outside world as the people in the Winlock area enjoyed.

The people of Lewis county as a whole were overwhelmingly white, with less than one percent recorded in the 1890 census as "Negro" or "Indian, Chinese, Japanese, and all other." Eighty-four percent were native-born, and over sixty-six percent were of native white parentage. In the study precincts, Germans and Scandinavians were the most numerous immigrant groups in both 1890 and 1900. The county's population as a whole more than doubled between 1900 and 1910, after which growth continued at a slower rate. In the precincts studied here, population numbered 4,251 in 1900 and 8,303 in 1920.[15]

Despite the trumpetings of its boosters, Lewis county retained many of its frontier characteristics in 1890. With a total land area of 2,336 square miles, population was still thinly spread. People were concentrated in the north-south corridor along the rail line, and less densely in an east-west band along the Cowlitz and Chehalis Rivers. Rough dirt roads, quickly transformed to quagmires by the northwest rains, were all that connected many of the tiny settlements and scattered farms to that beacon of civilization, the railroad. As late as 1920, men outnumbered women. The fertile prairies of the river valleys were quickly claimed by settlers, but the timberlands tended to remain vacant even after most of the trees had been sent to the mills. The enormous stumps of old-growth northwest forests required so much labor to clear that few settlers were attracted to the land, which turned out not to be particularly fertile anyway.[16] The depression which began in 1893 also dimmed prospects considerably, at least for the immediate future.

Although the economy of the area was already diverse, most people in the study area in the 1890s lived in farming households. According to the 1900 census, sixty percent of the households in the area, including the towns, were headed by farmers. In the more remote eastern precincts, farmers headed over ninety-five percent of house-

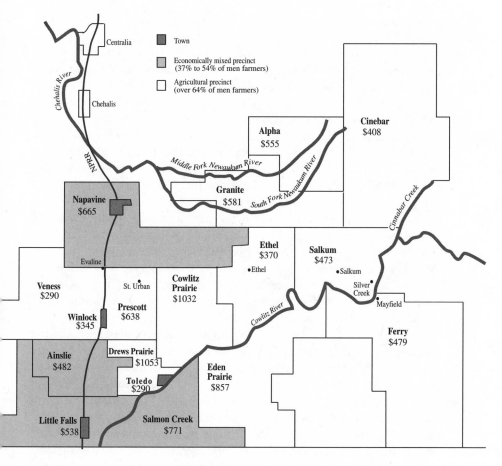

Map 2. Average real property values, 1900

holds, while the area along the rail corridor was more mixed (see Map 2). In several heavily forested precincts adjacent to the towns, over half the men engaged in logging or other occupations, but on the prairies of Cowlitz, Drews, and Eden precincts, over seventy percent of the men over age sixteen identified themselves primarily as farmers. Despite the depression and low prices for agricultural produce, new people continued to move to Lewis county during the 1890s and acquire farmland. Between 1890 and 1900, the number of farms in the county more than doubled. Most families were able to purchase their farms without going into debt during this period.

Over three-quarters of the farm families in the study area owned their land free and clear in 1900.[17]

The average Lewis county farm during the 1890s was moderately sized and supported a mixture of crops and animals. As population rose and the earliest settlers sold portions of their land or divided it among their children, average farm size declined from 160 acres in 1890 to 126 acres in 1900. About one-quarter of that farmland was improved in both years.[18] The typical family raised most of the food they and their animals consumed and sold what was left over. Hops was the only major crop raised primarily for market. In May 1898, Peter Sommersett reported on the number of acres devoted to different crops in Alpha precinct for a special edition of the county newspaper, the *People's Advocate*, assuring readers that, "I know every person in the precinct and know exactly what they have in the way of land, improvements and crop." He recorded 90 acres in wheat, 120 acres in peas, 500 in oats, 40 in rye, 600 in hay, 30 in potatoes, 40 in fruit, and 7 in hops. The same edition of the newspaper reported cattle, hogs, grain, hay, potatoes, apples, and hops had been shipped out of Chehalis the previous year.[19]

Farming in Lewis county, as throughout much of the United States, was a family affair. The census manuscripts, recorded by local residents following detailed standardized instructions, usually list only adult men and teenage boys as farm workers, but it is clear that women and younger children also did considerable amounts of farm work. Most families probably followed a fairly traditional division of labor. Throughout most of the United States, men were responsible for field work, including clearing, plowing, planting, harvesting, and marketing field crops. Women were in charge of the vegetable garden and poultry, preserved most of the food the family consumed throughout the year, and sold extra butter and eggs. Children helped their mothers, with boys assuming male tasks as they grew older. Care of the larger stock might be shared or differ from family to family.[20]

The ideal was not always practiced, however. In Lewis county, men often took paying jobs, leaving most of the farm work to their wives and children. For example, Ed and Clara Carpenter homesteaded near Toledo. Their son James later recalled that while Ed taught school, Clara cleared the fields and planted orchards. Joe and Maggie Ryan lived four miles east of Toledo, along Salmon Creek. Joe worked as a logger during the 1890s. Meanwhile, Maggie and the four of her seven children who survived infancy raised vegetables, cows, chickens, pigs, and turkeys. When money was tight, Maggie and her sons drove wagonloads of vegetables, butter, meat, and eggs to the mills to sell, while the girls "kept house." The 1900 census, however, lists Ed and Joe as farmers and Clara and Maggie as housewives.[21]

In the Hanken, Schultz, and Neels families, three German households with neighboring homesteads on Eden Prairie, the men also worked in town during the summers, leaving their wives and children on the farms. As the sons grew older, they too earned cash from off-farm labor. By 1900 John and Louisa Hanken, who had fourteen children altogether, had three sons in their twenties living at home and working in logging camps, while their seventeen year old daughter worked as a farm laborer. Neighbors Carl and Lettie Schultz also had a grown son working as a logger.[22]

Farming was perceived as a way of life, not strictly a business venture, even though families clearly intended to produce market crops for profit. Rural Americans shared many values with townspeople, including prevailing gender expectations which assigned public leadership and competitive roles to men, and emphasized motherhood, family duty, and virtue for women. But rural people also valued mutuality and cooperation within the family and among neighbors and took pride in the hard physical labor performed by both men and women on the farm.[23] In this latter respect, they believed themselves to be different from urban dwellers. Reminiscences of turn-of-the-century settlers in Lewis county are filled with descriptions of the hard work the women did on the farms, even when the men were not off working another job—stories clearly told with much pride.[24]

The economies of the towns depended both on the agricultural production of the surrounding countryside and the timber industry. In the towns of Winlock, Napavine, and Little Falls, between fifty and sixty percent of the men were employed as low-skilled laborers, many in logging camps and mills, some in railroad work, construction, or casual labor. An additional twenty percent or so were skilled workers, including engineers and foremen. Businessmen and professionals accounted for twelve to twenty percent of the men in these three towns.[25] Toledo, with less reliance on mills and more on agricultural trade, had far fewer workers and over thirty percent of its men were in business or professional occupations. The rural precincts of Prescott, Ainslie, Salmon Creek, and Veness also had logging camps or mills and included roughly equal numbers of farmers

Table 1. Occupations of men in the study precincts, 1900

Precinct/ town	Number of men	Occupation				
		Farmer[a]	Laborer[b]	Skilled laborer[c]	Business/ professional person[d]	Student/ retired
Eastern farming precincts						
Alpha	72	83.3%	8.3%	0%	5.6%	2.8%
Cinebar	56	66.1	17.9	7.1	0	8.9
Ethel	54	80.8	9.3	0	1.9	11.1
Ferry	27	88.9	3.7	0	3.7	3.7
Granite	31	64.5	3.2	0	3.2	29.0
Salkum	62	83.9	6.5	0	3.2	6.5
Western farming precincts						
Cowlitz	140	70.7	16.4	0	5.7	7.1
Drews Prairie	37	78.4	8.1	0	2.7	10.8
Eden	81	71.6	17.3	1.2	1.2	8.6
Prescott	82	65.9	19.5	4.9	1.2	8.5
Economically mixed precincts						
Ainslie	59	37.3	49.2	6.8	1.7	5.1
Little Falls	120	38.3	37.5	11.7	7.5	5.0
Napavine	206	45.1	28.2	15.0	10.2	1.5
Salmon Creek	139	53.2	28.1	6.5	5.8	6.5
Veness	43	41.9	41.9	11.6	2.3	2.3
Incorporated towns						
Toledo	106	19.8	20.8	21.7	31.1	6.6
Winlock	233	3.0	52.8	17.6	19.7	6.9

[a]Includes farm owners, renters, and farm laborers.

[b]Includes day, mill, logging, and railroad labor.

[c]Includes foremen, sawyers, engineers, millwrights, teamsters, carpenters, miners, painters, and tinners.

[d]Includes merchants, pharmacists, saloon keepers, hotel keepers, mill owners, photographers, editors, teachers, clergymen, physicians, lawyers, musicians, undertakers, and artisan proprietors such as shoemakers, barbers, and blacksmiths.

and other workers. See Table 1 for a detailed picture of employment in the study precincts at the turn of the century.

Most women operated within the family economy, but some opportunities for independent earning were available, particularly in or near the towns. About twenty percent of adult women were in business or professional occupations in Toledo and Little Falls in 1900, with fewer scattered through the other precincts. Most women earned outside income through fairly traditional female occupations, including schoolteacher, seamstress, and storekeeper. Although the timber industry employed only a few women directly as bookkeepers or camp cooks, many found work indirectly serving male workers as boardinghouse keepers, waitresses, and laundresses. There is no evidence in the census, newspapers, or other records that women in these small towns worked regularly as prostitutes. Apparently such services were only available in the larger towns.

The lumber mills of the small rail towns were by nature precarious businesses. The mills closed frequently due to bad weather, financial conditions, and fire. Business tended to be boom or bust, and competition was fierce. The Capitola mill had a fairly typical history. It opened in 1889 in Veness precinct near Winlock and shut down temporarily in August of 1893 along with most of the other mills in the county. The following summer a forest fire destroyed much of the standing timber on Capitola mill lands. In February 1896, local papers reported the Capitola had three big orders from the Northern Pacific Railroad and eastern companies, but by that summer, the mill was in receivership. It reopened the following year under the management of the Prescott and Veness families, who also owned Winlock's other mill.[26]

The mills of central Lewis county were primarily small, locally owned concerns during the 1890s. Three Summerville brothers owned two of Napavine's four mills. They had arrived in Lewis county from Illinois in the late 1880s with little money, and ten years later were among Napavine's biggest employers. Meanwhile, their parents had established a farm nearby. Brothers Carroll and Arthur Brown from Maine, who owned another Napavine mill, had also been accompanied west by their farmer parents. The fourth mill was owned by George McCoy, a young man from Wisconsin who lived in town with his wife and three small children.[27] In 1900, local mill owners began to feel pressure from Weyerhaeuser and other big timber tycoons from the Midwest who were beginning to buy up large tracts of timber land, but the bulk of Lewis county mills remained in local hands into the twentieth century.[28]

The mills did attract a work force of single men, but most Lewis county residents in the 1890s lived in family groups, including timber workers. About 66 percent of the people age sixteen or over in the study precincts in 1900 lived in nuclear families (including married couples and parents with unmarried children). Another 12 percent lived in extended families (with in-laws, grandchildren, or adult siblings, but no unrelated people), and 6 percent lived alone. Only about 7 percent boarded with non-family members, and most households that included unrelated boarders were headed by a married couple, often with children present. In Napavine precinct which included both the town and the surrounding agricultural land, seventy men (or one-third of men over the age of sixteen) were employed in the timber industry. Only ten of those were boarders, while thirty-seven were married household heads and an additional twenty lived with close family members. Eighty-five percent were native-born, and two-thirds had native-born parents—the same percentages as for the county as a whole. In Winlock, where town and precinct boundaries coincided, about half the men were laborers of some kind. Here, railroad workers frequently did live in households of single male boarders, but only nineteen men fell into this category. Mill workers and day laborers usually lived in family groups. On the predominantly agricultural Cowlitz Prairie, twenty-three men were recorded as timber workers in the 1900 census. Eighteen of them were sons living on the family farm.[29]

Thus, in the 1890s workers in the timber industry were not separated from other small-town and rural residents by clear class, ethnic, or lifestyle differences. For many men, work in the mills and logging camps was a way to earn the capital to improve the home farm or start a small business. The apparent lack of any labor organization among Lewis county timber workers during this period reinforces the view that they were integrated into the farming and business communities rather than separated from them by clear class differences.[30]

If Lewis county farm families suffered from that often discussed rural curse of isolation, it was not because of any lack of organizations to join. Kinship and neighborly connections were reinforced and augmented by an assortment of voluntary associations that linked people to others in their communities and across community boundaries. Some of these associations were highly structured and affiliated with national groups, for example the fraternal orders, churches, and the Grand Army of the Republic (GAR). Others, such as literary societies and singing clubs, were more casually organized on a year to year basis, although still following established norms and allowing people to reach beyond their daily group of contacts. Finally, parties and dances were more clearly direct outgrowths of kinship and neighborhood bonds, but again followed patterns recognized throughout the area. The lines between these three levels of organization often blurred as kin, friendship, and neighborhood ties overlapped with organizational membership.

Parties were the most basic kind of organized socializing. Lewis county newspapers are full of gossipy accounts of taffy pulls, dances, card parties, and other neighborhood entertainments. Many of these parties included midnight suppers and went on until the "wee hours." Families with teenage children appear to have been the most likely to sponsor parties. For example, Rosinda and Jacob Hovies of Alpha, with five young adult children at home at the time of the 1900 census, gave three private card and dancing parties in as many months during the winter of 1897–98. Several other public dances were held in the neighborhood during the same period. Cowlitz Prairie was a particularly sociable community, with five separate dances given during the month of February 1894. One of the parties reportedly drew one hundred people.[31] In many communities prosperous families built halls and regularly sponsored dances that were open to the public, hiring a band and director and charging an entrance fee of seventy-five cents or one dollar. The Hovies family had such a hall. So did the Myers, another Alpha family, Klaus Bezemer of Mayfield, Hugh Herren of Cowlitz, and the Summerville brothers of Napavine.[32]

Casual socializing in mixed-sex groups was both an outgrowth of, and helped to reinforce, mutual respect between men and women. Osterud, in her study of women and men in rural New York in the late nineteenth century, found that the sharing of both farm work and social lives built respect for women and their labor, and in part counteracted the legal, economic, and cultural inequality that women faced.[33] The kind of casual socializing that was evident in Lewis county suggests a similar ethic—one that recognized sex differences and allocated resources and duties on that basis, but at the same time reinforced shared ground between men and women and held in high regard the contributions of both. Intergenerational ties were also strengthened through community gatherings, integrating young people into the adult community and passing on values from one generation to the next. Newspaper accounts frequently make reference to the "young people" dancing far into the night, but people of different generations clearly mixed at these social functions.

Usually such occasions provided time for fun, courtship, and the reinforcement of community ties. Occasionally they also became arenas of conflict. In July and August of 1891, the *Nugget* published a series of letters from Alpha residents arguing with some bitterness over the events of one evening when two neighbors gave rival dances. The accounts of the correspondents vary so greatly that it is impossible today to tell what happened. Apparently no violence occurred, simply groups moving back and forth between the two sites, leading the hosts to feel that dancers were being "stolen." One of the hosts was charging a fee—a common and accepted practice— and the other reportedly provided not only the music, but supper and beer for free.[34] Whether other sources of animosity between the two families existed is unclear. The serving of beer at one of the dances might have been at least part of the reason for the conflict. Certainly, feelings varied widely among Lewis county residents on the propriety of alcohol consumption. Particularly among the Germans, beer was an accepted part of life. Many of the Protestant churches strongly denounced alcohol consumption, however, and the Masons and Oddfellows frowned upon it, although they were hesitant to prohibit it among their members. The Woman's Christian Temperance Union was active in both Napavine and Winlock, and a male group, the Murphy Temperance Association, also met in Winlock during the 1890s. The two dance sponsors on this particular occasion continued to be neighbors for many years. Both also became members of the same Farmers' Alliance local the year following the incident, suggesting they had found a way to get along.

In Napavine as well, alcohol might have been the issue that closed down these semipublic dances, at least temporarily. The initial conflict was not reported, but in April 1896 the *Chehalis Bee* noted that a "select dance party" was given for young people from surrounding communities, strictly by invitation to avoid "any undesirable party being present." The paper went on to say that this lifted a ban of several years on such events, indicating that behavior not considered appropriate by the general community had been taking place at earlier functions.[35]

Literary and debate societies and lyceums were also popular forms of community social and cultural life. These associations were rarely long lived. They were usually organized for a season, then disbanded till the next year. Thus in the same locality there could be differences from year to year. In Ferry, a literary society met every Friday in the spring of 1893 with Mrs. Cora Travis and Mrs. Jennie Kelly as president and secretary, respectively. The following January, however, a "reorganized" literary society had all male officers, Klaus Bezemer, Eugene Wright, and Arthur Bridges.[36] Whether membership in the two groups was also single-sex, we do not know, although they were not announced as such in the newspapers. The Cowlitz Literary and Debating Society organized in December of 1895 had men as president and vice president, David Motter and Charles Henriott, and women as secretary and assistant secretary, Manda House and May Motter, daughter of David. All were also active in the local Methodist church, and the Motters were mainstays of the Farmers' Alliance.[37] An Alpha lyceum which boasted fifty members in the summer of 1897 had all male officers and debaters. In most cases, the participants in these societies held a range of political affiliations, and discussions were intentionally nonpartisan. For example, the Alpha lyceum held a mock trial one week, and another debated whether more honor was due Washington or Columbus.[38]

The hops harvest provided another opportunity for informal socializing. Although it accounted for only a small percentage of cultivated acreage, hops were Lewis county's most visible specialized market crop, excluding lumber. Farmers throughout western Washington raised hops, particularly in Lewis and Pierce counties. Hop yards of almost any size could be profitable. While there were some large yards put in by the more prosperous and commercially oriented farmers, most were only a few acres. The crop was labor-intensive and required seasonally hired workers. When ripe, cones had to be picked quickly and then dried in specially built kilns before being marketed or stored.[39]

Hops picking was a major community event. Whole families turned out to pick, camping out on the hop farm. The work was not strenuous, and while men could earn higher wages in the logging camps and mills, the pay was considered good for women and children. Many local people joined in the harvest. Migrant workers, particularly Indians, also came for the season. Most years, pickers earned $1.00 per box, three boxes being a good day's work. When prices were particularly low, growers only offered $0.75, and in 1911 a group of pickers struck for $1.25, only to find themselves replaced.[40] Tensions that occasionally arose in the hops fields did not prevent Lewis county residents from generally considering hop-picking season an open-air party. The family apeal and often festive nature of hop picking are evident in Figures 3 and 4, where the workers have bedecked themselves with vines. Big dances often followed the successful completion of the harvest.[41] Disputes over pay rates were mitigated by the fact that many of the pickers were neighbors, and racial tensions were eased because the Indians who came to pick stayed only briefly and did not disrupt established community life. In 1897, the *People's Advocate* published this romantic ode to the hop harvest, likening the hop yard to a Utopian state:

> Had Edward Bellamy visited the hop fields of Lewis County, he might have witnessed the actual working of his ideal state in which all kinds and classes of people performed the same service and received credits of the same value. . . . There is perhaps no place under the shining canopies of heaven, where men—and women too, for that matter—meet upon such an equal footing as in the hop field. They are each and every one absolute Monarchs over their particular box. . . . Here we have no agitation for shorter hours because each picker may begin when he pleases and leave off when his own sweet will dictates. There is no trouble from strikes because work is at hand for all. The life in the field is a happy careless one—a true picture of modern Bohemianism.[42]

Formal organizations also provided important centers for social life. Churches were numerous, and often divided by ethnicity as well as beliefs and codes of conduct. During the 1890s, Winlock had Baptist, Methodist, Christian (Disciples of Christ), Christian Scientist, and Lutheran churches. Toledo had Baptist, Presbyterian, Congregational, and Methodist churches; Napavine had Baptist, Presbyterian, Catholic, and Methodist; Little Falls Evangelical and Catholic; additional Catholic churches were in the countryside, on Cowlitz Prairie and at St. Urban.[43] The less populated Salkum and Silver Creek area had far fewer churches. The people of Mayfield built a Methodist church with the help of the Board of Church Extension of the Methodist Episcopal Church in 1895, in what one correspondent described as a "heroic" effort, given the sparseness of the population and the distance from the rail line. Salkum was also a regular Methodist preaching station and had an organized church that met in either the Salkum or nearby Burnt Ridge schoolhouse. A Baptist church was built in Salkum in 1901 and a Presbyterian congregation met there during the 1910s.[44]

The churches were the major exception to the general integra-

Figure 3. Hop pickers at Betty family field near Toledo, Washington, ca. 1895. Photograph courtesy of Lewis County Historical Museum, Chehalis, Wash.

Figure 4. Hop-picking crew at the Salem Plant yard in Salkum, Washington, 1895. Photograph courtesy of Lewis County Historical Museum, Chehalis, Wash.

tive tendencies of local associations. Members of Winlock's Lutheran church were mostly German immigrant farmers and their children. Every Lutheran head of household who could be linked to a 1900 census record owned his home free of debt. Church services were conducted in German until the first World War, and when a large number of Finnish Lutherans moved to the area early in the twentieth century, they started their own congregation with a Finnish preacher coming up from Portland, rather than join the Germans.

Most of the members of Winlock's Baptist church, on the other hand, were native-born of native parents, lived in town, and worked as laborers or housekeepers. Many of them had southern roots. In keeping with their occupations, Baptists were the least likely of all church members to own property, and those who did own property on average had less of it. Methodist churches were more widely spread throughout the area than the other denominations, with itinerant preachers who could serve several preaching stations. In addition to churches in the town centers, the Methodists had congregations in the countryside. Their memberships were varied, mixing native-born and immigrant and different occupational groups. Just over half the men were farmers and most of the women were recorded by census takers as housewives, but Methodists also included mill laborers and foremen, craftsmen, merchants, and even a saloonkeeper. (For information on how membership in an organization correlated with occupation, see Table 2.) Unfortunately, records from the Catholic churches were not available to link member names with census or other records. Some former Hudson's Bay Company employees, many of whom were Catholic, had settled on Cowlitz Prairie. Many Germans immigrants were also Catholic, and Catholic parishes were organized where the German population was most heavily concentrated.

Churches were formal organizations with national affiliations that determined many of their policies. Although women comprised a majority of most congregations and were recognized as important contributors, their formal roles were circumscribed. Men held a monopoly on the important public office of ordained minister, as well as on many of the lay offices. Few of the early Lewis county churches survive today, and among those that do, records are far from complete. The Methodist churches have the best surviving records, with complete membership and officers lists from five of the churches. In these congregations, only men held the office of trustee, which entailed financial responsibilities,

Table 2. Occupations of men and women belonging to organizations, 1900

| | | Occupation | | | | |
	Number of men	Farmer[a]	Laborer[b]	Skilled laborer[c]	Business/ professional person[d]	Student/ retired
Churches						
Baptist	18	16.7%	55.6%	5.6%	11.1%	11.1%
Lutheran	21	52.4	19.0	4.8	19.0	4.8
Methodist	61	47.5	21.3	11.5	14.8	4.9
Lodges						
Masons	49	44.9	28.6	4.1	18.4	4.1
Oddfellows	20	25.0	10.0	15.0	45.0	5.0
Eastern Star	21	19.0	33.3	19.0	23.8	4.8
Rebekahs	25	24.0	24.0	24.0	24.0	4.0
Farmers' Alliance	23	73.9	4.3	0	17.4	4.3

| | | Occupation | | | |
Organization	Number of women	Farmer	Housekeeper	Business/ professional person	Student/ retired
Churches					
Baptist	20	0%	95.0%	0%	5.0%
Lutheran	17	0	88.2	5.9	5.9
Methodist	75	2.7	89.3	2.7	5.3
Lodges					
Eastern Star	30	3.3	83.3	6.7	6.7
Rebekahs	27	0	74.1	14.8	11.1
Farmers' Alliance	4	0	100	0	0

[a]Includes farm owners, renters, and farm laborers.

[b]Includes day, mill, logging, and railroad labor.

[c]Includes foremen, sawyers, engineers, millwrights, teamsters, carpenters, miners, painters, and tinners.

[d]Includes merchants, pharmacists, saloon keepers, hotel keepers, mill owners, photographers, editors, teachers, clergymen, physicians, lawyers, musicians, undertakers, and artisan proprietors such as shoemakers, barbers, and blacksmiths.

but both men and women were stewards, lay leaders, and sunday school superintendents.[45] In Winlock's St. Paul's Lutheran Church, with its largely German membership, the perceived differences between men and women were emphasized by having women sit on one side of the sanctuary, men on the other.[46]

In all of the churches, women had their own organizations. The

Ladies' Aid or Altar societies gave women the opportunity to gather regularly for Bible study, socializing, and money-raising activities. In addition to their weekly meetings, the church women's groups sponsored events for the whole community. Ice cream, strawberry, or taffy-pulling socials were frequent festivities. An Easter-time egg social given at the Eadonia or East Toledo Methodist church in 1893 attracted 150 people. Such activities enhanced the image of the congregation in the broader community and reinforced the ties and commitment of members. The money the women raised also helped furnish church kitchens, buy pianos, augment ministers' salaries, and sustain mission work.[47]

The bulk of Lewis county white settlement took place during the period of peak popularity for fraternal lodges in the United States.[48] Lodges were consequently among the early organizations founded. In the fall of 1891, T. F. Kennedy, William Connahan, and Klaus Bezemer of Ferry walked for miles searching for fellow Masons in the sparsely settled area so they could start a new lodge. They managed to find five others and so founded the Robert Morris lodge.[49] At about the same time a Masonic lodge and its women's auxiliary, the Order of the Eastern Star, were organized in Winlock, probably with greater ease. Toledo residents organized a Masons lodge in 1901 and the Eastern Star in 1919. The Robert Morris lodge of Ferry, which moved to Silver Creek in the early twentieth century, opened an Eastern Star lodge in 1910. Oddfellows lodges met in Winlock, Toledo, and Little Falls during the 1890s, and their women's auxiliary, the Rebekahs, in Winlock and Toledo. The Modern Woodmen and the Women of Woodcraft organized in Winlock and Napavine in 1899 and 1900, and the Woodmen of the World in Toledo at about the same time. The Grand Army of the Republic was a similar kind of association specifically for former Union soldiers. It also had a women's auxiliary, the Woman's Relief Corps, as well as a Sons of Veterans auxiliary, with lodges at Winlock and Toledo.

The lodges inspired great loyalty among their members, with their secret rituals, codes of conduct, and language of brother- and sisterhood. Members who lived in the countryside remembered walking along the railroad tracks or riding horses as far as fourteen miles to attend meetings, and spending the night in the homes of town members.[50] One of the major functions of the lodges was social. They frequently sponsored dances, masquerade balls, suppers, and basket socials, many of which were open to the public. The lodges consciously mixed generations, with mothers and daughters, fathers and sons joining together. Grown children and siblings living in separate households also joined the same lodges, strengthening both kinship ties and the sense of the lodge as an extended family. The fraternal groups provided benefits for their members and members' families for funerals and in times of sickness or impoverishment. The Toledo Rebekahs paid for new eyeglasses for one of their members in 1895 and made contributions to the Oddfellows home for elderly and destitute members. In 1908 the Toledo Masons were supporting four indigent members.[51]

The fraternal lodges had more internal diversity than the Lutheran and Baptist churches, but also differed from the general population. Overall, lodge members tended to be native-born, better off than average, and more likely to remain in the area through the 1910 census. Between ten percent and twenty-five percent of lodge members were immigrants, mostly Scandinavian, British, or German. Native-born lodge members came from throughout the United States. All lodges had a substantial minority of farmers, but town dwellers clearly predominated, in contrast to the population as a whole. About a quarter of male lodge members were in business and professional occupations and male lodge members were more likely than average to own both real and personal taxable property, as Table 3 shows. Lodges were clearly not, however, a bastion of the business elite. Unskilled and skilled laborers as well as farmers were well represented in the lodges. The vast majority of lodge women were recorded as housewives by census takers, but a few were occupied as teachers, dressmakers, or in other occupations.[52]

The lodges were organized formally by sex and family relationship. Only men could belong to the Masons or Oddfellows. Male lodge members and their female relatives could belong to the Eastern Star or Rebekahs. The Masons/Eastern Star were the most strict, but all the fraternal lodges were similar. A woman could join a church or the Farmers' Alliance on her own initiative. She could only join the Eastern Star, however, if sponsored by her husband or father who was already a Mason. Women's independence was further undermined by the non-reciprocal method of organization in the lodges. While the men's groups were exclusively male, the auxiliaries were mixed-sex. Offices within the auxiliaries were assigned by sex so that male domination remained secure. The Rebekahs and Eastern Star provided an arena for male lodge members to socialize along with their wives and daugh-

Table 3. Average property assessment and percentage of members holding taxable property, by organization, 1900

Organization	Real property		Personal property	
	Average assessment (in dollars)	% of adults assessed	Average assessment (in dollars)	% of adults assessed
Churches				
Baptist	376	13.2	61	23.7
Lutheran	408	31.6	188	34.2
Methodist	482	26.1	146	30.4
Lodges				
Masons	727	46.9	214	55.1
Eastern Star	884	15.7	139	25.5
Oddfellows	334	45.0	166	65.0
Rebekahs	548	23.1	204	25.0
Farmers' Alliance	786	66.7	296	59.3
Average all adults	620	22.7	179	26.0
All men	617	35.1	184	41.2
All farmers	673	59.7	169	63.4

ters, not an independent vehicle to give women their own voice in the orders. The auxiliaries also provided a handy pool of female labor to cook the meals and organize the dances. The Masons and Oddfellows on their own rarely sponsored any kind of social event, public or otherwise.

Fraternal lodges were phenomenally popular among both men and women during the second half of the nineteenth century. The roles of fraternal lodges in inculcating middle-class values to working-class men and both building and easing social barriers have long been noted by historians. Recent studies of nineteenth-century fraternalism have also begun to explore the gender implications of the lodges' ritual and organization. Mark C. Carnes believes lodge ritual provided a means of shaping and making sense of masculinity in the economic and ideological context of Victorian America. As virtue and religion became increasingly associated with the feminine, fraternal lodges provided an alternative, secular, and distinctively male approach to moral living.[53] In Mary Ann Clawson's analysis, women's auxiliaries were established by the lodges after the Civil War at least in part to help reconcile prevalent notions of gender, which assigned the keeping of morality and virtue to women, with the claims of male fraternal lodges that they held the key to moral living. But women were never allowed to become full members in the orders. Clawson explicitly links the bonding by men in their own exclusive organizations with the continued male monopoly on public power during the nineteenth century. The fraternal lodges certainly did not create male domination, but they helped sustain it.[54]

The lodges avoided any direct political discussions, allowing people of different political philosophies to meet comfortably. Only the political affiliations of those few lodge members who ran for elective office or were active in the party hierarchies could be identified, but even this limited data shows a fair mix. The Masons were almost equally divided, with five Republicans, three Democrats, and six Populists. In fact, Judd Herren, the president of the Lewis County Farmers' Alliance, was one of the charter members of the Toledo Masons, bringing his demit from a North Carolina lodge. But that lodge was formed only after the Farmers' Alliance and Populist party had faded. The five founders of Ferry's Masonic lodge which was active in the 1890s included a Republican and a Populist.[55] No Populists were found among the Oddfellows, but equal numbers of Democrats and Republicans—eight each. Two Alliance members were identified among Toledo's Rebekahs, one of whom left the lodge in 1895 as the result of an unspecified "trouble," and was then invited back with the promise that "she would be treated as one of the Sisters again." The wording of the secretary's minutes suggests that the "trouble" was personal in nature, not political.[56]

Other studies have found a similar blurring of class and political lines within fraternal lodges, even while they advocated an ideology generally associated with the middle class.[57] The class-oriented rhetoric of the People's party attracted many adherents in Lewis county during the 1890s. But no matter how heated the debate got in the newspapers and from speakers' platforms, Populists and Republicans, farmers and businessmen, timber cutters and mill owners still shared the language of brother- and sisterhood at their lodge meetings.

A letter in the *Chehalis Bee* makes clear the general belief that political discussions did not belong in certain arenas. In this case, the association was the Napavine Lyceum in 1897. One person sent a letter to the *Chehalis Bee* complaining about the trivial nature of the discussions and the lack of substantive political debate in the Lyceum. The following week a reply appeared: "Why did he not call . . . and make up his lost friendship? We think that would be more manly than to put such a foolish little article in print and have it circulating about the country." The letter concluded: "There are a few cranks that try to attend societies just with the hope of stirring up peaceful citizens."[58] Here, maintaining friendship despite political differences is viewed as "manly"—a term of praise—while a person who insists on discussing issues of disagreement in a primarily social setting is criticized for disrupting community harmony.

The blurring of class lines in lodges was made easier by the general fluidity of economic and social standing in a changing community. Status certainly differed by occupation, and people had varying levels of wealth, but class is difficult to discern in a society such as this. Most people engaged in physical labor of some sort, and many were able to accumulate property as fairly young adults. No one lived in splendor, although by the end of the decade some town businessmen were able to build large Victorian houses, and a handful of people had live-in servants.[59] Farmers and their sons frequently worked in the mills or logging camps to earn money to invest in their farms. People from various backgrounds could open a shop with only a small amount of capital. Clear lines between those who labored and those who profited from property, or among merchants, workers, and farmers, cannot be drawn in this period. With all the fluidity and murkiness surrounding class, the political differences expressed in the struggle between Populists and Republicans, discussed in the following chapters, primarily reflected divisions between town and countryside.[60] In this setting, the Farmers' Alliance, firmly rooted in the countryside, could be a vehicle for political expression without disrupting other community ties. The Masons and Oddfellows on the other hand, with their mixed rural and town membership, carefully avoided overt partisan discussions.

Other lines were not so easy for the majority to overcome. Race could be a factor in exclusion from community activities, although newspaper and other accounts present a complex view of racial attitudes, with a fair degree of fluidity here as well. Overt prejudice was apparently minimal. Lewis county's population was overwhelmingly white by the end of the century, but did include a few African Americans and Native Americans. In the study area, the 1900 census takers recorded seventeen adults as black or mulatto and seven as Indian. It is clear from other documentary evidence that some people described as mulatto in the census were of mixed white and Indian blood, and some families with some Indian ancestry were identified as white. Hudson's Bay Company men and other early settlers often married Indian women, and those who remained after Oregon and Washington territories became officially part of the United States were able to take advantage of generous land claim acts and become substantial property owners. Their children then intermarried with white newcomers and blended with the predominant Anglo-American culture. For example, Abel St. Germain's father was French Canadian and his mother the daughter of one of the area's first white settlers and an Indian woman. The 1900 census identifies both Abel and his brother as white, although their ancestry was widely known. When Abel married a young woman from Bohemia in 1897, a local newspaper reported that he was of French and Indian descent, and "well and favorably known" in the area.[61] Race was mentioned in the account, but was not apparently considered by the reporter to be a deterrent to the marriage or an issue in Abel's social standing.

Native Americans who maintained their Indian identity were also respected by whites. When the last hereditary chief of the Cowlitz Indians died in 1912, the *Nugget* gave him a glowing obituary and deplored the fact that the United States government had never rewarded the friendliness of his tribe with either money or their own reservation. Six years later, the Chehalis businessmen's club opened its meeting hall to over 200 Cowlitz Indians for a tribal convention.[62] At the same time, white residents never publicly questioned their right to occupy the Cowlitz's former homeland.

The small African American community that settled near Winlock around the turn of the century also generally lived at peace with their white neighbors. The first black family in the area, headed by Tom and Jane Lynch, arrived in 1892. Tom and Jane had been born in North Carolina like Judd and Susan Herren, and were also in their mid-40s when they arrived in Washington. As was common for residents of Lewis county, this was not the Lynch family's first move—their three children living at home in 1900 had been born in Tennessee. We can speculate that the Lynches moved on to Washington with the same aspirations of owning land and providing for their children that had motivated the Herrens and others. The Lynches bought a farm in Veness precinct which they owned free and clear. Their property was valued below the average of all farmers in the study area, but above the average for Veness. They joined Winlock's mostly white Methodist church, but also donated land for a black church—a kind of separatism that was also practiced by the Germans and later by the Finns. At the time of the 1900 census, their daughters attended school, presumably alongside their white neighbors. Small numbers of black children continued to attend the Veness school into the 1910s, as can be seen in

Figure 5. Students and teachers of Veness school, 1915. Photograph courtesy of Lewis County Historical Museum, Chehalis, Wash.

Figure 5. Although African Americans shared much in common with white residents of Lewis county, some distinctions were no doubt evident. Tom and Jane probably had been slaves until their late teens. They now owned property and participated in some aspects of public life, but their names do not appear on the roles of the fraternal lodges or in connection with the Farmers' Alliance. By 1910, Tom was a widower living alone on his farm. No trace of his daughters can be found in the census manuscripts at that time, but Tom's niece, Mary Whitesides, and her husband's family, also from North Carolina, continued to live in the area at least into the early 1920s. Andy Whitesides logged for the Veness Lumber Company, at times serving as crew boss, and also owned moderately valued property.[63]

As long as African and Native Americans acted in ways acceptable to their white neighbors, they were tolerated and even welcomed, but a less positive side of race relations is also documented in local newspapers. Those who stepped out of line were condemned not just as individuals, but also as representatives of their race. An *Advocate* report of a fight in a saloon in Little Falls contrasted the demeanor of a "drunken half breed" with that of an "industrious law abiding citizen"

—presumably white.[64] In 1894, the *Nugget* ran an editorial condemning the use of nonresident Indian and Chinese labor for hop picking. Ironically, the editor identified Native American as well as Chinese workers as alien, and concluded, "The hop pickers are as a type floating. We fondly hope and believe . . . that our home people are beginning more and more to receive the earnings of the hop picking season, rather than Indians or Chinese."[65] Similar fears of competition from Chinese labor were expressed in an 1891 editorial in the *Bee* protesting the opening of a Chinese laundry in Chehalis when there were white women wanting the work, and in this particularly vicious item that appeared in the *Nugget* in 1893: "About two thousand Chinamen were roasted to death last week in the burning of a Chinese city with an unpronounceable name. Of course this distressing mortality [sic], but since it had to be, we would have been much better satisfied if this respectable number had been suddenly subtracted from the pig tail population of this coast."[66]

Anti-Chinese prejudice was at a peak throughout the West during the late 1880s and 1890s, as the transcontinental railroads were completed and tens of thousands of workers, many of them Chinese, flooded the volatile labor market. Violence directed against Chinese workers occurred in several states in the 1880s. White mobs drove even long established Chinese residents out of Tacoma and Seattle. Fear of Chinese competition provided a major impetus for labor organization under the banner of white solidarity up and down the coast in the last years of the nineteenth century.[67] Even though the people of south-central Lewis county had little if any face-to-face contact with Chinese people during this period, they must have been affected by the blatant racism so casually expressed around them.

People were also well aware that prejudice and ill-treatment were common against African and Native Americans, and at times condemned such behavior. In 1897 the Newaukum Prairie Debating society, just northwest of Napavine, followed a debate on the Chinese Exclusion Act with the question whether Indians or Negroes had suffered more in the United States.[68] The incidents that reveal racial attitudes most clearly occurred in Adna, a few miles northwest of Napavine, and Chehalis in 1901. According to the *Nugget*, an unknown gentleman attended a masked ball at Adna. His "shapely form" and graceful dancing made him a great favorite of the ladies. When masks were removed at the end of the evening, he was found to be a member of the Siwash tribe. The article went on to recount a similar incident at a Chehalis cake walk, where a "darky boy" disguised as a white girl was the clear favorite of the crowd—until the disguise was removed. The article concluded with a pointed question: "would the applause have been half as hearty if the masks had been removed before the cake walk commenced?"[69]

In both of these incidents, it is clear that the assembled crowd assumed everyone in attendance was white. It is equally clear that no retributive action was taken or proposed against the interlopers. The newspaper reports the masquerades as pranks and takes a moralistic stance against prejudice. Apparently, most white Lewis county residents generally did not feel threatened by the small numbers of African Americans and Indians living in their midst and were willing enough to have them as neighbors, as long as they conformed to community standards. At the same time, prejudice barred African Americans and Native Americans from participation in most community organizations and events.

Closer to home than the virulent anti-Chinese agitation of the period was the anti-Catholic prejudice of the American Protective Association that surfaced in the mid-1890s and will be discussed in Chapter 4. A substantial minority of the county's population was Catholic. Although anti-Catholicism is an ongoing theme in northwest history from the 1840s, Lewis county newspaper reports indicate that Catholics, and particularly German Catholics, were in general highly regarded. Organizations and activities of German Catholics were advertised and noted favorably in all the newspapers, and items on funerals held at Catholic churches frequently mentioned the large attendance and high regard in which the deceased had been held. The *Advocate* declared in 1898, "No more patriotic people than the Germans exist, where the welfare of the American Union is at stake. . . . The future of the German race in this country is very promising."[70]

Religion, like race, played a role in lodge membership, but in this case it is more likely that people of certain religious affiliations chose not to join rather than that they were excluded. A number of practicing Methodists, Baptists, and Presbyterians were lodge members, as were immigrants from Scandinavia, Germany, and Ireland. Most Catholics, however, apparently followed the general condemnation of lodge membership of their church and did not join. German Lutherans were divided on the issue, to the point that a group broke from Winlock's St. Paul's Lutheran church in 1900 over the "Logenfrage"—the lodge question. They formed a new church, St. Peter's, which remained separate until 1970 when the two reunited.[71]

Dancing and other popular activities also divided church people. Baptist, Disciple, Presbyterian, and Methodist ministers of Lewis county gathered occasionally as the Ministers' Association of Lewis County. At an animated meeting in Centralia in February 1897 they condemned dancing, card playing, and the theater, all common entertainments in the lodges and in less formal gatherings. Apparently the ministers expected and welcomed a public debate on the issues, for they called an open meeting for the following week.[72]

The membership logs of the several Methodist churches include notes on the reasons for people withdrawing—usually because they left the area, sometimes because they went to another church. Sometimes people also left, or were asked to leave, because of conduct. "Unbrotherly conduct to pastor," "gross immorality," or simply "backslid" were among the notations. Several people also left Winlock Methodist Church because they refused to abstain from dancing, and one because she and her husband owned a saloon. Some of the women of the church had formed a Woman's Christian Temperance Union (WCTU) chapter, and she was tired of the opposition to her business.[73] Use of alcohol was also an issue in the lodges. In 1906 and 1908 the Toledo Masons put one of the brothers on trial for "unmason-like conduct" for selling alcoholic beverages. As the man was a bartender by profession, there was no question that he sold liquor. The issue was whether this constituted unmason-like conduct. After the first trial the charges were thrown out by a two-thirds majority, but on the second occasion, the bartender was ejected from the lodge.[74]

The Robert Morris Masonic lodge was split by a far different question, one that affected the whole community. The founders had originally named Ferry as the home of the lodge. At the time, Ferry was a boom town of some promise with a sawmill, hotel, and several shops. For the next several years, however, residents of Ferry and Mayfield, only one-half mile away, competed over which would predominate. Location of the post office was a major point of disagreement. Rival applications for postmaster were consistently filed, with Mayfield residents declaring they would rather walk to Silver Creek for their mail than to Ferry.[75] Despite the turmoil, the Robert Morris lodge prospered and reportedly attracted over two hundred people for the officers' installation ceremonies in both 1895 and 1896. In 1895, Mayfield won the struggle as Ferry "blew up," and the lodge moved its headquarters to Mayfield.

This did not, however, end the local conflict. By 1905, Silver Creek residents felt the lodge should move closer to them. Mayfield, too, had declined. The 1904 Polk Directory of Lewis County lists Mayfield as only a "country post office," while Silver Creek is distinguished as a village. The dispute became bitter enough that the Grand Lodge, the state-level organization, almost withdrew the Robert Morris charter, but a compromise was reached. The lodge remained in Mayfield until the death of the town's founder and one of the order's mainstays, H. C. Mayfield, then moved to Silver Creek where it has remained.[76]

Community life in south-central Lewis county during the 1890s was complex. People of different ethnicities, economic standing, religions, moral beliefs, and political philosophies lived in close proximity to each other. The many voluntary associations gave neighbors a place to meet and common cultural forms that could overcome many of their divisions. Within the churches and lodges, people of different interests, beliefs, and ages mingled and called each other brother and sister. As in biological families, the resulting ties were sometimes ones that might not otherwise have been chosen, but they were apparently strong. Conflicts certainly occurred, and breaches were sometimes long lasting. But the overall ethic that inspired the associations was one that valued unity and commitment to the community. Sometimes beliefs were too strongly held to be silenced, but usually if people in an organization could not agree, they simply ignored the issue and stayed on safe common ground.

The variety of organizations also provided community residents with a means to express conflicting values without personally attacking their neighbors. One could join a church that frowned on drinking and dancing, thereby taking a public stand, and then simply ignore local festivities. The very diversity of the area worked against any clear divisions. People who differed on one issue were likely to find themselves allied on another. There were too many crosscutting lines for any of them to be unbreachable. Most of the voluntary associations in south-central Lewis county were apparently fairly open. Members did not conform to a narrowly defined norm. The Lutheran church and the Farmers' Alliance, discussed in the next chapter, were the most homogeneous. In the other churches and lodges, immigrants and native-born, farmers, businessmen, day laborers, housewives, and employed women seem to have mixed freely. Potential lines of conflict between town and countryside, business and working class were smoothed over.

The population of south-central Lewis county was diverse. The people who moved to the area in the last decades of the nineteenth century came from all over North America and northwestern Europe. There were pockets where Germans or Scandinavians predominated, but every precinct contained people born in many different places. There were also clear economic distinctions between town and rural residents, and between business owners and wage laborers. Community organizations thus had the potential either to integrate the population and provide a way to overcome—or at least manage —differences, or alternatively, to segregate and further emphasize differences in background and beliefs.[77] The evidence here indicates that in this early period community associations, particularly the fraternal lodges, tended more to integrate than divide. Despite the clear differences and divisions, most long-time residents of the area shared a commitment to family and community prosperity, and believed that working together was the surest way to achieve both.

✢ 3 ✢

New Visions: Political Culture
in the Farmers' Alliance

J. M. Smith, the Washington state organizer for the Farmers' Alliance, met with farmers near Boistfort in June of 1891 to form the first Lewis county local. Although some county newspapers ridiculed the Alliance and its aims, farmers throughout the county were reportedly eager for Smith to organize in their neighborhoods too. By August when Smith moved on to other parts of the state, Lewis county had eleven locals with a total of 300 members, and a County Alliance organization with father and son, A. J. (Judd) and Hugh Herren from Cowlitz Prairie's Cyclone Alliance, as president and secretary, respectively. The Alliance kept growing. By the quarterly county meeting held at Toledo on January 5, 1892, Lewis county boasted seventeen locals, and by the next year twenty-four.[1]

The Farmers' Alliance added a new dimension to rural public life. The Alliance was rooted in existing community values and followed many of the familiar organizational structures of other voluntary associations, but by combining social activities, economic self-help, community improvement, and political education into a single organization, it created new vision and energy. In the Alliances men and women, teenagers and grandparents met together in neighborhood groups to talk, sing, eat, and dance. At the same time, the Alliance connected Lewis county residents to a broadly based political reform movement through traveling lecturers, numerous publications, and participation in state and national meetings. The political culture that flourished in this open community base enabled Alliance members to transcend politics as usual and bring new issues to the political agenda. Goodwyn has described the Alliance as a "free social space" from which arose the democratic spirit of the Populist movement.[2]

Less well recognized by historians is that the Alliance also provided a free space for women to emerge as political partners with men. The combination of the ideological commitment to equality and justice and the organizational inclusion of all family members led to a transcendence of the gender norms that usually limited women's political participation in mixed-sex settings. Some historians have noted routine exclusion of women from Alliance business, particularly in the South.[3] But in Lewis county, women received the same education as men in the Alliance and came to be seen as not only equally capable, but equally responsible for political activity.

In the 1890s in Lewis county and most of the rest of the country only men voted, ran for office, and participated in party conventions. Women in the towns of Lewis county had access to the political arena through female associations that lobbied local governments on "municipal housekeeping" issues, as did urban women throughout the country. For example, the St. Helens Club of Chehalis was organized primarily as a social and educational club, in which middle-class women could study art, literature, and other topics. They also pushed for civic improvements and charitable causes. Women of Winlock raised the money to construct and run an orphanage in the 1890s, and in later decades organized a club devoted to improving their town. In the early twentieth century, Toledo women organized a Civic Club that cleaned up vacant lots, furnished parks, and lobbied the town council for sidewalks. Little Falls and Napavine women had similar clubs.[4]

Farm women, however, rarely had the leisure or the physical proximity to participate in such female associations. Their social life revolved around mixed-sex activities, and the entrance of the Farmers' Alliance in the 1890s gave farm women of Lewis county a chance to participate in the "male" world of electoral politics. Men and women did participate in many activities of the Farmers' Alliance in gendered ways, that is, they tended to perform those functions that were considered most appropriate to their sex. Men assumed the leadership roles in the locals and ran the county-level organization, while women provided food and entertainment. But both studied political economy, spoke on the "issues of the day," and helped spread the Alliance message. Furthermore, as in other realms of life, women were understood to be capable of performing "male" tasks and were expected to do so when the need arose.

Certainly the absence of women in the higher levels of party and political organization as the Alliance merged into the People's party illustrates the disparity in power between men and women in the public arena. The persistence of that disparity one hundred years later underlines its significance. Enduring differences between men and women at higher levels of the organization, however, make the extent and enthusiasm of women's participation at the community level all the more interesting. If political activism was a new departure for many rural women, involvement in the public life of their communities was not. They were the mainstay of churches, teachers and activists in the local schools, members and participants in fraternal lodges, officers of the literary societies. These institutions formed the core of community social and public life, and in Lewis county they existed before the Farmers' Alliance was organized. Had the focus of the Alliance been exclusively on "male" concerns of economic cooperation and electoral politics, women's participation would have been doubtful. The emphasis on social activities in the Alliance, however, made it only a small stretch for women to be involved here too, even though farm economics and political issues were central concerns of the organization.[5]

The Alliance grew out of a number of farmers' associations and political movements in the South and Midwest during the 1870s and 1880s. Farmers faced tremendous difficulties in the emerging industrial economy, particularly those in the recently settled Plains states and the war-ravaged South. Many were forced to take out loans to improve their farms and sustain their families. At the same time, worldwide agricultural production was increasing as new land was opened to farming, transportation improvements made much of the world accessible to the international market, and European powers encouraged the raising of staple crops in their new colonies. Wheat and cotton prices fell steadily as supplies increased, but American farmers continued to raise ever larger crops in order to pay back their loans. Further, as farm income fell, more and more farmers were forced to take out loans to keep going until the next crop, adding to the cycle. Deflation exacerbated the problem. In the years following the Civil War, the money supply remained tied to the gold standard. Population and both industrial and agricultural output increased enormously, but the supply of gold and thus of money did not, lowering commodity prices still further. While farmers suffered, industry seemed to benefit from a number of governmental favors, including high tariffs on manufactured goods, generous subsidies to railroad companies, and suppression of labor unrest. Most railroads, too, enjoyed monopolies on service to particular parts of the country and were not averse to price gouging. Farmers who depended on the market for their living were incensed by the control that creditors, distributors, and railroad companies exercised over them, and insisted that corporations and individuals must act for the common good as well as for profit.

Farmers throughout the South and Midwest organized around these issues. The Grange, or Patrons of Husbandry, initially showed the most strength. Its founders developed a ritual patterned after the immensely popular Masons but broke with the fraternal lodges in including economic and political concerns and admitting women as equal members. Enthusiastic recruits forced the enactment of railroad regulation in some midwestern states in the 1870s, in what are known as the Granger laws. The Grange spread as far west as Washington territory, including a local on Cowlitz Prairie. Grange membership plummeted, however, after 1875 when initial hopes for political and economic reform were not realized, and more conservative leaders took over the national organization. In some areas including Washington, the Grange died out completely, at least temporarily. Over the next several decades, other farmer organizations took the lead in championing political and economic reforms that would rein in the unrestrained greed of monopoly capitalism.[6]

By the mid-1880s, the Farmers' Alliance had emerged from its Texas origins and was becoming the most prominent farmer organization. The Alliance press and traveling lecturers preached a complex message, educating rural people in political economy and advocating a more just and democratic society. The Alliance sought government control of the money supply and the railroads for the common good, rather than private control for individual profit. It proposed an innovative subtreasury plan of government-owned storage facilities for nonperishable produce and low-interest loans to farmers, that would allow farmers to hold back their crops when prices were low and sell when prices were high. It organized buying and selling cooperatives in an effort to break monopolies and increase farm income and sought improvements in numerous services to rural residents. It demanded an end to special financial interests influencing the government, and increases in direct democracy through such means as the initiative and referendum, direct election of senators, the secret ballot, and in some areas, women's suffrage. By the end of the 1880s, the Southern Farmers' Alliance was well established throughout the South and Southwest. A second Alliance organization, the National or Northwest Alliance, flourished in the Midwest. Leaders of the Alliances and several other reform groups, frustrated by their inability to wrest economic power from large corporations or break into the political system, founded the People's party in 1891. The third party ran well in areas of Alliance strength during the 1892 elections. With the onset of a deep depression in 1893, its reform agenda became a major challenge to the two established parties.[7]

Alliance organizers began work in Washington state in 1888. They established numerous lively locals in the wheat-growing regions east of the Cascade Mountains, where a single crop economy, dependence on rail freight service, anger at the large government-granted land holdings of railroad companies, and a high rate of mortgaged indebtedness made farmers highly receptive to the Alliance message. The Grange regained a foothold in Washington the following year, championing a similar program of reform. But through the 1890s, the Washington Grange remained small and committed to nonpartisanship, with most locals clustered near the Oregon border and none in Lewis county. It was the more dynamic Alliance, with an aggressive national political agenda, that won widespread support from Washington farmers.[8]

The Northwest Alliance was the first of the two national-level Alliances to organize in the state. By 1891 when Lewis county organization began, the Southern Alliance, or the Farmers' Alliance and Industrial Union (FA&IU), which was the stronger on the national level, was challenging the dominance of the Northwest Alliance in Washington. According to Fred R. Yoder, a rural sociologist who studied the movement early in the twentieth century, the Northwest Alliance made a big push to organize Lewis county, and the local newspaper referred to Smith, the organizer, as an agent of the "National Alliance." A number of the locals, however, referred to themselves as members of the FA&IU, and the Populist newspaper established by Alliance members in Chehalis in 1892, the *People's Advocate*, identified itself as an official organ of the FA&IU. No discussion of rival affiliation appears in the Lewis county newspapers. Yoder claims that most Washington state members wished to see a unified organization, and the State Alliance repeatedly requested either joint membership in both Alliance organizations or a unified national body. These requests were denied. At any rate, Lewis county Alliance members do not appear to have made an issue of any rivalry that might have existed, and clearly believed themselves to be members of a nationwide union of farmers.[9]

By the mid-1890s, most Alliance locals in the state and around the country had abandoned any pretense of nonpartisanship and openly endorsed the Populist party. Although at first an organizational advantage, the virtual unification of the Alliance and the Populist party contributed substantially to the rapid decline of the movement when the national party organization disintegrated after the 1896 election. The Grange was to outlive the Alliance and ultimately become the dominant farmers' association in Washington and Lewis county early in the twentieth century.

The Lewis county residents who participated in the Farmers' Alliance appear to have been fairly typical of the rural population, although the historical record here is spotty. As in most parts of the country, Alliance records for Lewis county have not survived. Local newspaper accounts are the only record we have of Lewis county Alliance activities and members. Alliance members identified in the newspapers were most often officers, committee members, or in leadership positions of some other kind, and we would expect them to be among the better established members of their communities.[10] However, while members of the organizations discussed in the previous chapter were much more likely than average to remain in the area from one census to the next, Alliance members actually moved away more frequently than the average, so that the names that could be linked to census and tax records were relatively few. Most of this group were native-born, middle-aged, married male farmers, who owned their land free of debt. The average Lewis county farmer who was head of a household in the period would meet the same description, but the Alliance leaders owned more property than the average farmer (see Table 3). These Alliance leaders were also less likely to be in debt, more likely to be married, and had more people than average living in their households. The handful of known Alliance members not listed as farmers or housewives in the 1900 census included two ministers, a teacher, a merchant, and a day laborer. Only one of the identified Alliance members in the study area was foreign-born, Theodore Myers, president of Alpha Alliance, who had immigrated from Germany in the 1860s and was one of the most prominent farmers on Alpha Prairie. Women were probably undercounted here, since women rarely served as officers or committee members, and name changes upon marriage make women more difficult to track through multiple sources. Of all the names found of Alliance members, twenty-two percent were women; of the twenty-seven names that appeared in census records, only four were women.

At first glance, Lewis county farmers as a whole seem poor candidates to participate in the great Populist crusade. Few were trapped by debt, and most were able to grow a variety of crops and at least feed their families, easing the effects of low crop prices. While most historians have emphasized the economic distress that led farmers to organize and rebel through the Alliance, many also acknowledge that it was not those at the bottom but those with something to lose who became most active.[11] Many Lewis county farmers felt that debt and the loss of their farms were real threats. Most had migrated within the past ten years and had spent their savings on moving their families and establishing a new farm. They had not yet had time to accumulate much cushion against a series of bad years. Farmers feared losing all despite their hard work and questioned the justness of a system where hard productive labor did not result in a comfortable living. They were practiced in banding together with their neighbors to do cooperatively what could not be done alone and saw in the Alliance the potential to change things for the better. In 1892, a Salkum resident wrote to the *People's Advocate* that the Farmers' Alliance was growing steadily because so many feared going into debt and losing their farms unless government policy were changed: "The people have settled right down to consider seriously how many land owners will become tenant farmers before the next general election rolls around, and unless the government comes to the rescue by issuing money direct, instead of through National banks, many who own land today will soon be among the renters and paupers."[12]

The Farmers' Alliance consciously fostered a mixture of social and political life, and so combined community-building activities, education, and public involvement. The organization had a dual purpose: to uplift the farm family through social contacts and cultural enrichment, and to improve the farmers' economic lot through economic cooperation and political activism. The same local Alliance might organize a community Christmas tree, sponsor a literary society, and hold a picnic rally at which Populist candidates spoke.

Among the most active locals in Lewis county was the Toledo Telephone Alliance. It was formed by national organizer Smith during his first swing through the county in June of 1891. Sixteen people signed on as charter members at that first meeting. Thereafter, the local grew rapidly and steadily through 1896, then faded from existence, along with the rest of the Alliance organization in the county. In 1893, the Telephone local correspondent to the *Advocate* described the typical Alliance meeting, saying, "We talk of hard times, sing songs, make funny speeches, and have a good time in general."[13] Having a good time was certainly an important part of the Alliance's appeal. Dinners and dances were given frequently, sometimes just for Alliance members and guests, sometimes for the general public as fund-raising and recruitment events. In 1893, the Telephone local gave at least two "grand balls," one in April and one in October, and a dance and supper in July. All were termed great successes by the *Advocate* correspondent.[14]

Dances were not unique to the Alliance. Individuals frequently gave dancing parties, and many organizations sponsored "grand balls." Such events were among the few sources of entertainment available. They also brought people together in important community-building ways. Groups of Alliance members, or Oddfellows, or whatever organization was sponsoring a ball, would travel from nearby communities to share in the gala event and help cement their common ties. At the same time, balls allowed neighbors belonging to different fraternal organizations and political parties to reach out to each other in a way that emphasized their shared cultural values.[15]

Picnics and rallies were events that performed similar community-building functions, and at the same time helped spread the political message of the Alliance. Within a few months of its forming, the County Alliance sponsored a "grand picnic" on Cowlitz Prairie at which "noted" speakers both entertained and educated in Alliance principles. Farmers from all over the county came with their families and baskets of food.[16] Republicans and Prohibitionists held similar picnics as well. Rival party members came to heckle—as well as join in the fun.

The Alliance also helped organize Fourth of July celebrations, although these events were designed to overcome political divisions and emphasize the unity of all Americans. Businessmen's committees appear to have been the traditional organizers of July Fourth doings. Chehalis and Centralia committees often cooperated so that the two nearby towns would not have competing celebrations. Big crowds were an important element. Games, races, prizes, declamations, fireworks, and dances were all standard parts of the all-day events. Few towns or organizations had the energy to put on such a big show every year, so they took turns. In 1894 people of Alpha spent the Fourth at Morton, 23 miles to the east, with a day of sports, literary entertainment, and dancing. Three years later, they reciprocated and organized a celebration attended by people from Morton, Burnt Ridge, Cinebar, and Agate.[17]

In 1892, the County Alliance announced it would organize the celebration in Chehalis. Sam Herren, the Populist leader regularly savaged by the opposition newspapers, was to be a speaker, but both the *Chehalis Bee* and the *Chehalis Nugget* put aside whatever partisan differences they might have felt and boosted the event with front page headlines, just as they did for businessmen organizers. Any antirailroad sentiment among Alliance members was also set aside: excursion trains with low rates were arranged for the day so that people could come from all over the county—or at least from places near the rail corridor.[18] The papers reported the day was quite a success, although the weather was cool. John Alexander, Jr., whose mother was a member of the Urquart family, prominent in Chehalis and Napavine business and county Republican circles, wrote his father, "Had a big time on the Fourth the farmers alliance had charge of the celebration." He added, "They are going to make an interesting election this fall."[19] Alliance locals, too, sometimes organized July Fourth celebrations. The Telephone Alliance took on the task for the Toledo area in 1895, and Alliances of Alpha, Salkum, Ferry, and Harmony helped the Silver Creek local organize a celebration in 1893 at which Judd Herren spoke on Populist party principles.[20]

The organizers of these big public events are rarely listed in any surviving records. When the newspapers did print a list of organizers, they were almost always all male. Women often were included on literary programs and sometimes gave speeches. They certainly prepared the food that was an important part of the proceedings. But there was something distinctively male about the Fourth of July celebrations. The games and races which were advertised in the papers with their prizes and sponsors were for boys and young men. Participation by a young woman in these sporting events would surely have been unthinkable. Perhaps the self-conscious inclusiveness of the celebrations, crossing neighborhood and organizational boundaries, also helped push them across the line from a community event in which women would have unhesitatingly taken central roles, to a "public" event, like a party convention, that was "naturally" male-dominated.[21]

Within the Alliance itself, women's contributions were far more visible. Organizing the social events and hospitality that were such an important part of the Alliance's appeal was understood to be the women's responsibility. Women made coffee and cakes, sang songs and recited poetry, prepared food baskets for the sick, and helped organize public social events for their neighborhood locals. They performed similar tasks for the regular quarterly meetings of the County Alliance as well. Almost every write-up on the county meetings mentions the "bounteous" feasts prepared by "the ladies." Organizing food and lodging for the county meetings required considerable logistical skills and a lot of work. Delegates frequently spent several nights, given the transportation difficulties. At the July 1894 quarterly meeting held at Silver Creek, three hundred people attended and were well entertained. Meetings in the late fall and winter rainy season attracted far fewer. Only forty delegates showed up at the Winlock convention in January 1893 and twenty at Harmony in October 1895. But the women still had to be ready to provide for many more. In January 1896 the women of the Telephone local prepared a "sumptuous feast" for the county meeting, only to be disappointed by the low turnout. In this case, they blamed not the winter weather and roads, but the (male) officers' failure to publicize the meeting.[22]

Men, on the other hand, organized the business side of those quarterly meetings and occupied most of the offices in both the local and County Alliances. The information we have on officeholders is far from complete. Newspapers from time to time listed local officers, if the correspondent sent the names in. The *Advocate* printed Alliance directories, including the president and secretary of each local, but these lists are also incomplete. We can still get a general view of the extent to which men and women shared offices within the organization from the information available. Men clearly held the majority of all offices. In 1892 the *Advocate* printed the names of president and secretary of thirteen locals. The secretary of the Newaukum Prairie Alliance, between Napavine and Chehalis, was a woman, Mrs. Henderson. Captain Wuestney, later the *Advocate* editor, was president. The following year, of the eleven locals listed, only the Eden Prairie Alliance had a woman secretary, Mrs. Etta Murphy. For most of the locals we have no other listing of officers, but the Telephone correspondent did furnish additional information. Maggie Patterson served as vice president in 1892 and 1893, the highest recorded office held by a woman in the county, and Minnie Griffin was elected secretary in the fall of 1893.[23] Perhaps other women's names would turn up if we had complete officers lists. The office of lecturer was sometimes held by women in other parts of the country, and later in the Grange.[24] Nonetheless, it is clear that when it came to the formal position of power represented by office-holding, women were a tiny minority. Still, they were present. Barriers existed to full participation by women, but they could sometimes be breached. At the county level, the exclusion of women from formal positions of leadership was complete. No female officer of the County Alliance is ever listed in the newspapers. Lists of members "prominent in the work" and serving on committees were exclusively male, with a single exception: at the October 1895 convention in Harmony, Minnie Griffin served on the time and place committee.[25]

This division of labor neatly paralleled that existing throughout society. It is not at all surprising to learn that women assumed the duties of housewifery and men those of business and public leadership within mixed-sex associations. Part of the appeal of the Farmers' Alliance was that it reaffirmed the values of the communities it entered. It was an extension of life on the farm where both men and women had important but different duties that each could take pride in. Such a delegation of tasks violates our modern sense of the meaning of inclusion. In our view, the women were second-class members. I do not think the people of the 1890s viewed the situation the same way. Alliance women, by doing the "women's work" within their locals, were helping make a vital organization. They attracted new members and made the old want to keep coming. Food and funny songs were more of an inducement to attendance than speeches on the money question. But once at the meeting, both men and women learned about the financial and political issues as well. Then they were motivated to unite to lobby the legislature or form a cooperative or join the Populist party. The Ferry Alliance correspondent to the *Advocate* certainly viewed the gendered division of labor in this way. He proudly announced the local now had thirty-seven members—not bad for a sparsely settled area —and had good attendance at their weekly meetings, even in the winter. He concluded: "Thus the conquered heroes come to their conquerors, the heroines, who are literally loaded down with ammunition whenever an assault is to be made on the community, consisting of music vocal and instrumental, hot coffee, pie, cakes, etc., etc. This unexpected attack bewilders the enemy so that they fall an easy prey to their captors."[26]

This writer acknowledges the importance of women's contribution to both the Alliance and other community associations, suggesting that little organization would be present without women's collective initiative. While the imagery of women conquering the bewildered men plays with the stereotype of women as the weaker sex, this writer also fully accepts traditional gender roles. Women's contributions to political discussions and campaigning in the Alliance, on the other hand, clearly challenged the prevailing sexual division of labor. Even though women were primarily observers in public displays of citizenship like the July Fourth celebrations, within Alliance locals political education and activism were a family affair. Women turned out for political meetings and debates, helped organize picnics and rallies, and did their best to get out the vote by making election day a major event. They viewed political issues as of vital concern to themselves and their families, just as their husbands did. This statement does not imply that women were interested in politics only inasmuch as it impinged on the domestic sphere. On the contrary, many women were fiery orators, appealing to high ideals and national problems that went well beyond what could be construed as the traditional woman's realm. Both women and men were motivated to political activism by complex mixtures of self-interest and ideology. Women could see as well as their husbands that roads turned into muddy swamps with every rain, that crops cost more to produce than they could be sold for, and that the family did not have the cash to pay its taxes. Women were just as interested as the men in the root causes of these problems and the political solutions.

On March 1, 1895, the *Advocate* printed a notice of a meeting of the Telephone local the following Saturday that concluded, "All the members and especially the ladies are earnestly requested to be present." The next week, the Toledo correspondent reported the meeting had been well attended and lively, and that "the ladies were even more enthusiastic than the gents."[27] Inviting "especially the ladies" was common throughout Lewis county during the decade, and indeed, in many parts of the country in rural areas. Did this mean the women were less likely to attend meetings? Probably. Despite the prevalent mixing of sexes in community associations, perhaps the standard nineteenth-century ideology emphasizing the woman's place in the home still held enough power to make many women feel reluctant to attend unless specially invited. Perhaps, too, women were often inclined to stay home and get the children to bed, rather than bundle them all into the wagon and try to keep an eye on them during the meeting. After all, most women who had been married over one year could expect to have children to care for well into middle age. Participation in community public life did not make childcare any less a woman's job. Constant urging might well have been necessary to keep even committed members attending regularly. "Especially the ladies" might even have been standard code for "it's okay to bring the kids."[28]

Women showed their enthusiasm not only by turning out to meetings despite the difficulties, but by speaking as well. The Little Falls correspondent to the *Advocate* wrote just before the election in 1894, "One should hear how earnest the women are in our cause. Mrs. Knapp and Cauldron can discuss the cause of the people with a vim that will put to flight the most eloquent of the money bag party."[29] The major speaker at the County Alliance picnic at Toledo in August 1895 was a woman, Mary Hobart of Whatcom county in northwestern Washington, whose discussion of finances earned her the description, "one of America's brainiest women," from the *Advocate* reporter. At the same picnic, Edith McNulty, a member of the Telephone local and daughter of one of the County Alliance leaders, combined the more standard female role of literary entertainer with political exhortation by reciting a poem on money.[30]

In May 1895 the Telephone local announced it would discuss

woman's suffrage at the next meeting. The notice in the *Advocate* concluded, "The lady members should come prepared to defend their side of the question."[31] This announcement is particularly interesting. Underlying it are two clear assumptions: that women would participate in the discussion, and that men and women would tend to have different views on the subject. Woman's suffrage is viewed here as a women's issue, not an Alliance issue. Unfortunately we have no record of the arguments expressed at this meeting. Some local Alliancemen and male Populists were strong advocates of women's suffrage. The *Advocate*, always under male editorship even with frequent changes, was a consistent supporter of the cause. In November 1894, it ran a prominent article headlined, "Women Must Vote Now!", concluding, "Women will vote, and some day we may have a woman president when the people come in."[32]

Countering this image of relative equality in the give and take at the locals are other suggestions that women were routinely excluded from some Alliance business meetings. In newspaper announcements of formal debates to be held at Alliance meetings, the debaters listed were almost always male. In one instance, the correspondent to the *Advocate* from the Centralia local wrote of a particularly enthusiastic meeting in 1895 that included a welcome surprise. He said, "At the call 'good of the order' a rap was heard at the door, which proved to be a call to supper prepared by our sweethearts and wives." He went on to praise the food and entertainment.[33] The implication here is that the men were at the meeting while the women prepared the food, and the men noticed nothing unusual in the women's absence during the discussion of Alliance business. Further, the women were described by their relationships to the men present, rather than as members in their own right.[34]

Party politics were never far below the surface in Alliance meetings. Entertainment, food, and dance might have been prominent features of the meetings, but so was examining the "issues of the day." The Telephone local regularly held meetings open to the community featuring staged debates by Alliance members. Two of their topics in late 1895 were whether senators should be elected by direct vote, and whether laws regarding money, rather than money itself, were the root of all evil.[35] Regulation of money was at the heart of the Populist agenda, and direct election of senators was one of its elements for expanding democracy. Such debates were both a widely used form of entertainment and a means of educating people on political issues.

Partisan politics came to dominate the Farmers' Alliance by the mid-1890s. A. J. and Hugh Herren, leaders in the County Alliance, were also prominent in the local Populist party organization, and the *People's Advocate* was clearly the voice for both the Alliance and the Populist party. In Toledo's Telephone Alliance, a new recruit was identified as a "former Republican," making clear the connection in at least some people's minds between joining the Alliance and changing political affiliation. Alliance correspondents from Cinebar and Silver Creek wrote expressing similar assumptions. Not everyone in the Alliance gave up their earlier party affiliations, however, and local people realized that partisanship could split the Alliance. At the fall quarterly meeting of 1892, A. J. Herren made a plea for brotherly unity and nonpartisanship, along with use of the ballot box to restore freedoms. Some locals took seriously the Farmers' Alliance claim of nonpartisanship and resisted the movement to merge the Alliance and People's party. The Winlock correspondent wrote the *Advocate* in 1892, stating the Alliance had a "grander object than partisan politics"—to right wrongs and bring harmony—and would outlive the political parties.[36] The Napavine correspondent, too, emphasized the need for harmony and cooperation of all farmers regardless of politics, while also making clear the connection between economic distress and subscription to Populist principles: "Let us not antagonize the wealthier class of farmers but take pains to point out to them, the advantages of belonging to the Order, as it [the Alliance] should have the cooperation of the entire farming population."[37] Some Alliance members did leave, evidently, as the Alliance merged into the People's party. Theodore Myers was president of the Alpha Prairie Alliance in 1892, but is not mentioned as active in the organization in later years, and served as a delegate to the county Republican convention in 1896. Still, political reform had been a clear goal of the Alliance from its inception in Lewis county, and for most Alliance members, the Populist party was the obvious route to reform. The correspondent of the Lamb local near Napavine expressed what may have been the prevailing view of nonpartisanship. He wrote they rarely discussed politics, since the order was supposed to be nonpartisan, but added, "We are all of the same opinion"—presumably Populist.[38]

Political education was both a collective and family affair. It was assumed, or at least hoped by activists, that children of Alliance families would be taught reform principles and grow up to vote for the Populist party. Birth announcements in the correspondents' columns of the *Advocate* often proudly proclaimed the arrival of a new Populist. One father wrote in, "Yes, it is a girl, though I would much rather it was a boy, for I do like to see the Populist party grow."[39] This father apparently did not foresee any changes in political gender roles that would bring women the vote, or perhaps was merely reflecting the barriers women faced in moving from the Alliance, with its mix of social and political, to party politics. Still, the clear message of the *Advocate* overall was that men, women, and children could all contribute to the cause of reform if they were knowledgeable about political economy. Alliance activists saw a direct line between knowledge and action. Just after the 1892 election, an *Advocate* editorial encouraged farm families to study together, and made clear the direct connection to political activism:

> Much may be accomplished between now and spring by reading up on the financial, economic and industrial questions. Let a whole family— father, mother, sons and daughters—set aside two or three hours each evening for reading and discussion. If tried where it has not been, the result will be surprising before the winter is gone. Farmers' clubs have been a great incentive in this direction, and laborers' clubs in the towns and cities may be made just as beneficial. And by all means let the women be interested in the work. The farmers' wives who have gone into the alliance work have been a wonderful help. That has been manifested by the interest they have shown in the recent political campaign. . . . Every man and woman in this country should be a politician in the true sense. Ours should be in fact as it is in name a people's government.[40]

People's party clubs, also known as Silver or Initiative and Referendum clubs, formed the direct link between the Alliance and People's party. Many Alliance members respected their order's official nonpartisan stance and overcame it by forming a separate organization during the campaign season, with separate officers and different meeting times. Almost every community had these clubs. References to Populist clubs are most numerous, but Republican McKinley clubs existed as well. The associations met weekly or biweekly to hear speeches and debates on political issues. Without membership lists, it is impossible to know the extent of overlap between the People's party clubs and the

Farmers' Alliance, but it was no doubt great. In June 1894, Cowlitz Prairie's Cyclone Alliance met to organize a People's party club. Hugh Herren, secretary of both his local and the County Alliance, was elected president of the new Cowlitz People's party club.[41] The Salkum People's party club alternated biweekly meeting sites between Salkum and Burnt Ridge, just to the north, in the summer of 1894. In June, one of their speakers was the deputy lecturer of the County Alliance. Of course, townspeople who would not have belonged to the Alliance could join the party clubs. One of the Winlock People's party club officers in 1894 was an Alliance member, but the other three officers whose names are listed do not appear under any Alliance activities.[42]

Although organized for explicitly partisan reasons, the party clubs did not exclude women. Given the rarity of female officers among the Alliance locals, it is not surprising to find only men among the club officers whose names appeared in the newspapers. A number of references make it clear, however, that women attended meetings and joined the clubs. The Salkum correspondent wrote, "The ladies of Salkum are certainly anxious to learn the cause of the hard times, judging by the number present. After the meeting I heard some promise to join the club first chance."[43] The casualness with which women were accepted as proponents of political parties is revealed by a news report in the *Advocate* in the fall of 1896. In describing a Republican rally the reporter stated, "men and women from all parties" were present. Other examples, however, point to female exclusion. The Toledo correspondent reported with some pride the organization of a People's party club in March 1894 with thirty-six members, "most of them being qualified voters"—in other words, male citizens.[44]

Apparently, women could be identified as Republican or Democrat as well as Populist. They packed the picnic lunches for those rallies, too, and no doubt wrote for and helped edit the partisan newspapers. The Ladies' Relief Corps of Toledo provided dinner for a Republican rally in October 1896.[45] There is little evidence to suggest, however, that women were active in Republican and Democratic politics to the extent that they were in the Populist movement. The very fact that *Advocate* correspondents continually called attention to and boasted about the high level of female participation and interest suggests that this was a departure from the political norm. The Populist party grew out of a community-based voluntary association that included men, women, and children. Without this base in the Alliance, it is doubtful that women would have had any more access to the Populist party than they did to local Republican or Democratic party organizations. Furthermore, the ideology and program of the Alliance and People's party consciously transcended politics as usual and pointed to a different and more inclusive vision of society that could embrace new gender norms, whatever the intentions of the organizations' founders. The high level of rural women's involvement in partisan politics in the 1890s was based on their active role in community life but was also a new departure.

Despite the challenges posed by these new visions, members of the Farmers' Alliance and Populist party were able to live for the most part peacefully with their Republican neighbors, in part because the organizations used the cultural forms that were already widely accepted by the community and through which people of different political beliefs could find common ground. At times, however, the conflicts broke out in violent and malicious behavior. The Farmers' Alliance was not just another fraternal order competing for members, but an organization that was challenging much of the existing economic and political order. The Telephone Alliance of Toledo was one of the most lively in the county. It was also centered in a town that remained solidly Republican at the height of the Populist movement. Perhaps open conflict in such a setting was inevitable.

In March of 1893 the Telephone Alliance proudly announced it had secured a lot, put up a fence, and would build a hall that summer. For every organization during this period, from Masons and Oddfellows to churches, putting up a building was a sign of having secured a solid place in the community. The Alliance gave a ball and supper as a fundraising event for their new hall. Two weeks later the posts for the hall were erected. Then during the night, a band of "outlaws" tore up the fence and posts and stacked the lumber in a mock monument over the supposed corpse of the Alliance. The Telephone local correspondent to the *Advocate* claimed the vandals were "known 'stalwarts' of the G.O.P.," and the same people who had been calling the Farmers' Alliance outlaws and anarchists.[46]

Apparently, the Telephone local did not get their hall built, whether because of the resistance they encountered or a lack of funds, we do not know. The following year the Alliance members fixed up an unused store building in Toledo for their meetings. This did not end the troubles. The next year, in February 1895, someone again tore up the fence from the Alliance grounds and carted away the boards. This time, there was no suspect, but the correspondent expressed hope that whoever had been responsible would return the boards.[47]

Populist electoral strength was rooted in the Farmers' Alliance. The Alliance took advantage of the existing gendered division of labor in community life to build a lively organization. The people of Lewis county understood the work of men and women to be complementary, and the contributions of both to be necessary because they were different. The celebrated culture of Populism, which flourished not only in western Washington but throughout the nation's periphery, relied equally on the work of women and men, and can only be understood by examining its gendered basis. At the same time, the Alliance drew rural women into political discussions and campaigning alongside men in a new way. Alliance women rarely spoke directly to "women's issues," focusing instead on the same issues of economic distress and political democracy that roused their brothers in the movement. Together, men and women of the Alliance were among the pioneers of a mixed-sex political culture that was to become the norm with the passage of the Nineteenth Amendment a quarter of a century later.

❦ 4 ❧

Populists and Republicans:
National Parties and Local Issues

Running its first slate of candidates in the general election of 1892, the Populist party swept many of the rural precincts in Lewis county. The Republicans held on to their lead in the towns and so were able to continue their domination of local government, but the Democrats were so disheartened by the enthusiasm of the Populists and the meager response to their own candidates that they sought a merger with the new party. The rank-and-file Populist converts spurned that offer, despite the urgings of their leaders. For the rest of the decade, the People's party posed a serious challenge to the traditional two parties in Washington and in many other western and southern states.[1]

Lewis county farmers were attracted to the Populists by the same issues that drew their counterparts elsewhere. They envisioned a cooperative commonwealth where producers owned the fruits of their labor and justice prevailed. They wanted government controlled by the people as a whole rather than by entrenched financial interests; monetary policies that favored debtors rather than creditors; government control of transportation and marketing facilities; higher prices for their crops; and lower taxes. At the same time, local issues and personalities played major roles in shaping and deciding the outcomes of political debates. The desire for good roads and economic development united Populists and Republicans who were sharply divided on national and state issues, while the jockeying of different factions for control of local party structures disrupted party unity. The surge and rapid decline of Populist strength was a national phenomenon during the 1890s, but one that was played out in particular local settings. Lewis county may not have been an entirely "typical" Populist stronghold, but it illustrates how local and national issues and people interacted in the building and dissolution of a movement.

In Lewis county, as in much of Washington and the northern United States in the decades following the Civil War, Republicans won most elections. That dominance met regular challenges by radical third parties, however, beginning with the Populists in the 1890s and continuing until the Democratic resurgence of the 1930s. Most of the challenge occurred in the countryside. The Republicans carried Lewis county even in 1896, the peak year nationally and in Washington state for the People's party, but only because of unflagging support from the towns. As early as 1892, before the onset of the 1893 depression that is often given credit for much of Populist strength, the farming precincts were returning sizable majorities for the Populists.

The Populist party had been born from the political union of numerous separate reform movements, most prominently the two Farmers' Alliance organizations, but also including the Knights of Labor, leaders of the Woman's Christian Temperance Union, and followers of Edward Bellamy's Nationalism and Henry George's Single Tax scheme.[2] In St. Louis in February 1892, an enthusiastic convention of reform organizations adapted the Alliance's platform to serve a new third party. The revised platform included a stirring preamble written by Ignatius Donnelly that proclaimed "the fruits of the toil of millions are boldly stolen to build up colossal fortunes," and concluded, "we seek to restore the government of the republic to the hands of the 'plain people.' " The assembly concluded with a call for the various organizations to choose delegates for a nominating convention of a new People's party.[3]

The extent of grass-roots enthusiasm for the new party was evident in the speed with which supporters organized their individual states, counties, and precincts. Lewis county, with an Alliance organization only a year old, ran a full Populist slate in 1892 that dominated the rural precincts. The Populist coalition was strong enough in Washington to elect their own John R. Rogers to the governor's seat in 1896, and a Populist-Democratic-Silver Republican majority to the 1897 state legislature. Other Alliance states in the South and Midwest showed similar victories. Nationally, however, the picture was not so bright for the reformers. The Populists failed to gain a major following in cities or in any part of the Northeast, leading many to question the ultimate viability of the party. In a compromise with political expediency, the Populists nominated liberal Democratic candidate William Jennings Bryan to head their own presidential ticket in 1896, and submitted to an emphasis on the remonetarization of silver to the exclusion of other important Populist concerns. A major rift resulted between those who favored fusion with the Democrats and those who sought to "keep to the middle of the road." Bryan's loss only deepened the divisions. The beginnings of recovery from the long depression, and the expansion of the money supply with the discovery of gold in the Klondike further dampened reform sentiment. While the national party crumbled in acrimony, farmers drifted away from the Alliance. Whereas the Democratic party had all but disappeared in much of Washington and other northern Populist states in 1896 and 1898, by 1900 it had once again replaced the Populist party as the main challenger to the Republicans.[4]

Although national economic concerns drew many to the party, the Lewis county Populist organization concentrated on local solutions to farmers' financial woes. County finances were one of the dominant issues of the decade's elections—less public expenditure would mean lower taxes. The first campaign platform of Lewis county Populists, passed in October 1892, attacked the county commissioners for collecting excess tax money and not initiating improvements in roads and schools, and for spending too much money going after delinquent taxpayers. The platform also echoed a concern already voiced by the County Alliance that the state tax law overassessed farmers and underassessed other property owners.[5] A running theme in every campaign year was the amount of county debt and what were considered excessively high salaries paid county officers. County debt was rising, from $5,545 in 1893 to over $23,000 in 1895. According to the Populists and their organ, the *People's Advocate*, this benefited not the honest working people of Lewis county, but politicians and bankers of the old parties out to defraud the people. Just before the election of 1894, the *Advocate* ran a front-page article under the headline, "Vote for Economy by Voting the People's Party Ticket." The article posed the question: "Shall two banks elect the treasurer of this county and manipulate its funds, or shall the people say who the treasurer is to be, where the money is to be kept and where the interest earned on it shall go? That is a pertinent question. Elect honest Tom Spooner, the farmer and you have solved the problem."[6]

After a victory in the auditor's race in 1894, the Populists could back up their claim that their party would save taxpayers money. In August 1895 the *Advocate* proudly compared the cost of running the auditor's office for the first six months of a Populist administration to the cost in the last six months of a Republican's term. According to their analysis, Populist Schooley had cost the county only $885.48, compared to $2,381.65 for the Republican. The article made no comparison, however, of the work accomplished or normally performed during those time periods.[7]

Many county officers, including auditor, sheriff, and treasurer, were allowed by law to hire deputies to assist with the work, and to charge the deputies' salaries to the county. Populist candidates for county offices were required by the 1894 convention to pledge to pay any deputies they hired out of their own salaries to save taxpayers' money and to prove their motivation for seeking public office was not their own gain. The Republicans countered by promising to reduce county officials' salaries by one-third. In January of 1895, a local judge ruled both these campaign pledges "reprehensible and illegal" attempts to bribe the voters, but this ruling did not stop similar pledges in following years. After the 1896 election, salaries were in fact decreased from $1,900 and $1,700 to $1,500 for all county offices. When the auditor, sheriff, and treasurer each requested $50 per month to pay deputies, the *Advocate* roundly condemned them for breaking their campaign pledge. A Napavine correspondent called Populist Auditor Schooley a "traitor," and stated, "There are many children obliged to go barefoot in Lewis county this wintry weather, on account of financial distress—yet some of our county officers possess the impudence to tell the county that $75.00 a month isn't enough for them."[8] Although the other county papers supported the hiring of deputies with county funds, Schooley bowed to the storm of protest and withdrew his request, sending his letter of apology to the *Advocate.*[9]

Well aware of the strength of the economic argument for joining the People's party, local Republicans responded by incorporating some of the Populists' concerns into their own platform. Attention to county officials' salaries was one example. The 1896 Lewis county Republican platform also called for no peacetime bonding of the federal government and waffled on the gold standard. They supported a "sound" currency, whether based on gold, silver, or paper, straddling the line between their own party's insistence on the gold standard as the only sound basis for the economy and the popular Populist-Democratic position that the money supply must be expanded if all were to share in prosperity.[10]

Collecting taxes from the Northern Pacific Railroad Company was another issue on which most county residents agreed, and where local Republicans chose the benefits to their community over their party's general support of corporate interests. When the Supreme Court ruled in 1896 that railroad companies had to pay taxes on their federal land grants, the county's dominant Republican newspaper, the *Chehalis Bee*, ran a long article under the headline, "Good News This." The county's attorneys had already been pursuing the Northern Pacific for two years, claiming the company owed Lewis county $64,000 in taxes. Over the next two years, local judges and county officials negotiated with the NPRR. Finally a settlement was reached in May of 1898 under which the county would receive $42,000. All parties hailed this as a wise decision that would end years of wrangling and bring the county some much needed cash.[11]

With the election of McKinley in 1896, Lewis county Republicans pronounced a growing prosperity. In August 1897, the Cowlitz Prairie correspondent to the *Bee* wrote that crops looked better than they had for years, "and this with a knowledge that an era of prosperity is being built upon a solid foundation and not a transitory one," made farmers happy. A few weeks later the *Winlock Pilot* declared, "The farmer will have about 16 dollars this year where he had one a year ago. This is not the kind of 16 to 1 a good many of them voted for, but they will probably be perfectly satisfied just the same."[12] Prosperity might have been more evident in the towns, however, where the mills were once again running at full capacity, than in the countryside. In 1898, a Salkum resident still spoke of hard times, writing that half the buildings in Salkum were vacant and that the people there had not "felt the McKinley boom yet."[13] Not until the summer of 1899 did the *Advocate* declare the beginnings of an economic turnaround. A front-page article in July announced that while for the past five years land could not be sold at any price in the county, timber lands were beginning to sell at appraised values, and some demand for farmland was being generated by renewed immigration. Booming mills, large crops, and Northern

Pacific track improvements were all creating a strong demand for labor and good wages.[14]

Fusion became an issue in Lewis county almost as soon as the People's party formed. The Democrats started out in second place and saw joining forces with the Populists as their only means of defeating the powerful Republicans, possibly even of surviving. In May 1892, the county Democratic convention attracted delegates from only eight of the county's twenty-seven precincts. In the meantime, the newly formed Populist party was booming, claiming one thousand members. At an enthusiastic nominating convention in August, the Populists rejected offers by the Democrats to create a unified ticket to run against the Republicans. At their own convention in Centralia later that month, however, the Democrats decided on a joint ticket, apparently with the agreement of the Populist county leadership.[15] The Republican newspaper, the *Bee*, claimed that Sam Herren, a former state legislator, chair of the state People's party committee, and a prominent Lewis county politician, as well as brother of A. J. Herren, county Farmers' Alliance president, was being bribed to bring off a fusion deal. This charge can be dismissed as an attempt by the partisan press to smear a leader in an opposition party. It does seem clear, however, that Herren, and probably others in the county Populist leadership, favored fusion from the beginning and pushed strongly for it. Herren was a town lawyer and an experienced politician, used to negotiation and compromise and yielding to political expediency in order to gain a greater goal. He was also a Southerner whose political roots had probably been in the Democratic party, although he served as a delegate to the Republican state convention in 1890 after his move west.

Fusion was not, however, popular with the more idealistic rank-and-file Farmers' Alliance membership, who insisted on "keeping to the middle of the road" and avoiding collaboration with either old party. In September, most of the Populist ticket resigned in response to the apparent fusion deal, along with John Nelson, chair of the county committee. The resulting uproar among local Populists forced the county leadership to call a new convention. This convention, too, must have rejected pressure for fusion, for in October the Democrats and Populists had separate county tickets.[16]

In March of 1893, the correspondent to the *Advocate* from the Cinebar Farmers' Alliance reported that their local was holding its own. On the fusion issue, the correspondent was unequivocal. "It is an easy matter to keep in the middle of the road at Cinebar," he reported.[17] The middle-of-the-road militancy and loyalty to the full range of Populist reform remained strong in the outlying eastern precincts, even as fusion and a single emphasis on the silver issue became reality at the national and state levels. In 1894 the Silver Creek correspondent responded to the debate going on within the People's party by stating, "It seems that some alliance men want to sidetrack the peoples' path on the single silver issue. That is not what we want. We want reform from the president on down."[18]

Fusion and the extent of involvement of the Farmers' Alliance in party politics were not the only issues to divide Alliance members and Populists in Lewis county. The *People's Advocate*, probably the most visible symbol locally of the reform movement, was shaken by a series of staff changes. The paper was owned by a large group of stockholders, including most of the leadership of the County Alliance. Several locals also owned stock. The overall editorial stance of the paper as an organ of the Farmers' Alliance and People's party was never questioned, but who would exercise direct control and determine specific editorial positions was very much an issue. A number of changes in editorship occurred. The first were relatively peaceful, made for personal rather than ideological reasons. Then in 1893 Charles Wuestney was appointed editor by the trustees. Wuestney had been born in Germany, moved to the United States as a young man, served as a captain in the Union army, and settled on a farm near Napavine in 1885. He had been active in the Alliance and People's party from the beginning and was one of the founders and trustees of the newspaper. His wife, Anna Hunsenstein Wuestney, was also a capable writer and frequently contributed to the paper. Anna Wuestney's tributes to the virtues of cool beer and Charles's reviews of the quality of brew served at local taverns might have alienated some in a reform movement that often included prohibitionists, but the Wuestneys balanced their views by providing weekly space to the WCTU.[19]

Wuestney's management was short-lived. In August 1894 the trustees abruptly dismissed him and took over editorship of the paper themselves. Walter Nevil, J. Kaylor, and S. M. Dunn claimed that Wuestney had been uncooperative, had refused to publish notices of stockholders meetings or to hand over the books, and the stockholders had shown their lack of confidence in him by rejecting him as a trustee in the last stockholders' election. Wuestney countered with a letter published in the Republican *Bee*. He claimed that Sam Herren and Judge Tugwell were determined to control the Populist party in Lewis county for their own benefit, and when he would not go along, they arranged a secret meeting with only a handful of selected stockholders present to elect new trustees and eject Wuestney.[20] The fact that Wuestney was an ardent middle-of-the-roader while Herren and Tugwell were reputed to support fusion suggests that party direction was very much an issue in the dispute. Wuestney remained active and popular in local Populist circles. He was invited frequently to speak to groups and was elected delegate to county and state conventions. At his death from heart disease in 1898, he was serving as treasurer of the county People's party central committee. By that time, however, one of his main rivals for influence, Sam Herren, had moved to Idaho, and fusion was a forgone conclusion.[21]

In June 1895 a long letter from Wuestney was published on the front page of the *Advocate*, urging Populists to keep to the middle of the road and to continue supporting the reform press. He concluded, "My private affairs stand away behind the reform movement."[22] Despite such moves for public unity, Wuestney's private affairs continued very much a public matter, and fodder for ridicule by the competing partisan press. Anna Wuestney instituted a suit against the *Advocate* and its stockholders, claiming that she was owed $1,000 plus interest for a loan she had made the paper in 1892 but had not been recorded in the books, and $440 in wages for the time she and her husband had run the paper from September 1893 to August 1894. The Wuestneys intended to hold onto the paper's books until the suit was settled.

It wasn't until June 1898, several months after Charles Wuestney's death, that the case finally came to a jury trial. The principals in the suit all agreed that Charles Wuestney had been entrusted with the contributions of the stockholders and empowered to run the paper, taking any profits for his own livelihood. The trustees denied Anna Wuestney had made the paper any loan, though she might have invested in it along with many others, and they claimed that Charles Wuestney had entered into the agreement on behalf of the marital community of himself and his wife, nullifying any claims for wages she might have over and above the papers' proceeds. The fact that there were no proceeds, that everyone knew the paper was a "losing proposition," was beside the point. Hugh Herren, County Alliance secretary and nephew of Sam Herren, was one of the witnesses, apparently in behalf of Anna Wuestney. The jury could not reach a verdict, so a new trial was ordered but never held. Evidently, the parties reached an out-of-court settlement.[23]

The *Nugget* had some sympathy for the Wuestneys in what it deemed their fight against "the lawyer gang" of Tugwell and Sam Herren, but the *Bee* commented with barely restrained glee: "The Populist leaders in Lewis County would have the general public to believe they are statesmen broad enough to run the state and county governments yet have fully demonstrated in this instance that they cannot run a little one horse newspaper successfully."[24] The fight for control of the *Advocate* reflected the larger battles that divided the party, as different individuals and groups with separate interests and opinions struggled to control official party direction. The Populists were not the only party to be affected by such controversies. In 1896 the entrance of the anti-Catholic American Protective Association (APA) onto the local political scene brought disruption to the Republicans as well.

The APA had been founded in the Midwest in 1887, during a period of high immigration, intense labor unrest, and highly visible expansion of Catholic influence nationally with the building of parochial schools and the success of Irish politicians in major cities. The organization sought to protect what it saw as "true" Americanism (that is, Protestant, Anglo-Saxon Americanism) from the influences of those who had a higher loyalty to a foreign ecclesiastical power (the pope) than to the United States government. The APA was a secret society whose founders had borrowed heavily from Masonic ritual. Unlike the Masons, however, it had overt political goals: to prevent Catholics from gaining power in politics or in that key institution of Americanization, the public schools. The new society grew only modestly at first, but the depression in the 1890s that helped fuel the Populist crusade also greatly increased anti-immigrant sentiment, as cities filled with the unemployed. In 1893 the APA spread to California, and in early 1894 lodges opened in Washington's major cities. The APA achieved modest political success in both Seattle and Tacoma, working through the Republican party in city council and school elections.[25]

The history of anti-Catholicism in the Northwest dates back to the 1840s, when the first Protestant missionaries to Oregon territory denounced the priests with whom they competed for conversions.

However, the evidence from Lewis county up until this time suggests an overall climate of religious tolerance. The Catholic church had long been prominent. The St. Francis Mission on Cowlitz Prairie was among the first institutions established by whites in what was to become Washington. Catholic French Canadians often held relatively high status by virtue of being among the first white settlers, and many of the German, Swiss, and Austrian immigrants establishing farms in the county in the 1880s and 1890s were also of the Catholic faith. While anti-Chinese racism was obvious in the local press, up until this time northern European immigrants, whether Catholic or Protestant, had been generally considered solid, industrious citizens.[26]

The first evidence of APA activity in the county appeared in 1894 just before the November election, when the Democratic *Nugget* strongly denied rumors, spread by Republicans, it claimed, that several candidates on the Democratic ticket were APA members. The following summer, the Democrats took the offensive. The *Nugget* ran a front-page article stating that the APA was gaining strength in Centralia and several other towns, and noting with some satisfaction that most of the recruits were Republicans. The *Nugget*'s editor clearly disapproved of the organization's objectives, and doubted the APA would make much further headway in Lewis county. The article emphasized the constitutional guarantee of freedom of religion, and concluded that as there were "many feisty" Catholics in the county, the APA was likely to face a good fight.[27]

In May 1896, when county conventions were held to elect delegates to the state party conventions, the APA began to show its clout. Ten of the thirteen county delegates elected to the Republican state convention were reputedly APA members, among them two Winlock businessmen. Not all Republicans approved of the society's influence. An editorial in the Republican *Bee* dolefully predicted APA domination would lead to defeat of the party's ticket in November. Talk of anti-Catholicism and the pope were inappropriate in county politics, the editor stated, and the majority of good Republicans who were not APA members would not stand for a secret society dictating their political candidates. The *Advocate* reported on a similar attempt by APA members from Centralia, Winlock, and Toledo to control the Populist county convention in early June, but claimed the attempt failed.[28]

Then on September 4 the *Bee* printed a front-page expose of what it termed the APA's efforts to control the Republican party and all other parties in Lewis county. Most citizens now had no voice in the political process, the *Bee* charged. Control was exercised by a handful of men who led the APA, including Republicans C. T. Hall of Winlock, John Ferrier, and Francis Thorne (editor of the *Toledo Tidings*) of Toledo, and prominent Populist A. P. Tugwell. To back up its allegations, the *Bee* printed a letter from Wilson Brooks, local head of the APA, urging all members, regardless of political affiliation, to register and vote in their parties' upcoming primaries. They could be sure that all Catholics were registered and planning to vote, Brooks claimed, so they must stick together, too.[29]

Over the next several weeks, the *Bee* continued to give prominent coverage to its story, emphasizing the efforts of "good" Republicans to regain control of their party in time for the November election. It also printed Wilson Brooks's response to the first article. Brooks defended the APA as a friend of all true Americans, committed to ending rule of the "Old Monopoly", and analogous to the Farmers' Alliance and People's party in its efforts to organize for necessary reforms. In fact, Brooks claimed, many of its platform planks were quite close to those of the Populists, a statement that must have made the *Bee* happy to print the letter. The *Nugget* and the *Advocate*, in the meantime, proudly proclaimed the independence of their own parties from APA domination, and ignored Brooks's assertions of shared principles.[30]

In the coming months, the APA sought to regain respectability by acting more openly and forming an advisory board comprising equal numbers of Republicans and Populists, but the negative publicity apparently ended the APA as a significant organization in Lewis county. On November 13, the *Bee* commented that only a faithful few in Centralia, Chehalis, and Winlock had voted as recommended by the APA leadership.[31] Thereafter, few references to the organization appeared in any of the papers. Nationally, too, the APA faded rapidly. The society's national leadership divided bitterly over whether to continue their allegiance to the Republican party during the 1896 elections or to try to increase their influence by taking a nonpartisan stance. Just as in Lewis county, no party wanted to risk alienating the increasingly numerous and powerful immigrant vote by aligning too closely with an organization like the APA.[32]

What effect, if any, the controversy had on the outcome of local elections is difficult to assess, but the limited evidence suggests that the influence was minimal. Republicans and Populists split the offices between them in the general election. Republicans did well in the towns, Populists in the countryside, as was to be expected given previous local trends and the general outcome nationally. While nativism and anti-Catholicism are serious issues in nineteenth-century American history, in the rural and small-town economy of Lewis county in the 1890s few people responded to the ugly lures of the APA. Catholics and Protestants lived too closely together and shared experiences and values that were too similar to perceive each other as serious threats. The fact that it was the Republican *Bee* that published the expose of APA attempts at influence, suggests that the issue was similar to the Wuestney scandal for the Populists: primarily a matter of individuals and factions jockeying for control of the local party apparatus.[33]

Voting in the study area reflected the pattern of surge and rapid decline in Populist strength exhibited elsewhere. In 1892, the Republicans retained a comfortable plurality with 43.4 percent of the vote in the presidential race. The Populists made a remarkable showing for a new party, however, capturing 24 percent of the vote countywide and actually beating the other two parties in thirteen of the county's thirty-five precincts. Of the precincts surrounding Winlock and Toledo, only Cowlitz Prairie and Napavine favored the Populists in that first election. In the less populous and heavily agricultural precincts to the east, however, the Populists won absolute majorities and earned over 80 percent of the vote in Alpha and Cinebar. See Table 4 for a tabulation of voting by precinct for 1892 and subsequent elections.

By 1896, the People's party was polling majorities in all the agricultural precincts but continuing to lose in the towns. Populist strength remained highest in the more remote eastern precincts. With the exception of Ferry, the eastern precincts also were more heavily agricultural, had more real property owners, and had a greater degree of residential stability than the Winlock and Toledo area precincts.[34] Farmers in the outlying area were also more likely to own their farms free and clear, but had lower average property values than farmers in the agricultural precincts near the rail line. Low property values did not lead directly to Populist votes, however. In fact in the Winlock area, the precincts with the highest average real property values also had the highest percentages of Populist votes (see Table 5 and Map 3). The degree of commitment to ag-

Table 4. Percentage of vote cast for presidential candidates of major parties in 1892, 1896, 1900

Precinct/town	1892[a]			1896[a]		1900[a]	
	Republican	Democrat	Populist	Republican	Democrat/Populist	Republican	Democrat
Eastern farming precincts							
Alpha	10.2	5.1	84.7	32.7	67.3	38.5	61.5
Cinebar	11.1	8.3	80.6	22.2	77.8	45.5	54.5
Ethel						35.9	64.1
Ferry	41.2	3.9	52.9	25.0	75.0	69.6	30.4
Granite				22.6	77.4	63.2	36.8
Salkum	20.3	24.3	55.4	21.9	78.1	47.4	52.6
Western farming precincts							
Cowlitz	24.0	32.7	43.3	38.3	61.7	46.9	53.1
Drews Prairie				33.3	66.6	40.0	60.0
Eden	40.5	10.8	48.6	44.6	55.5	59.3	40.7
Prescott	35.8	35.8	28.4	39.5	60.5	45.6	54.4
Economically mixed precincts							
Ainslie	62.8	34.0	3.2	72.0	28.0	77.4	22.6
Little Falls	45.2	48.4	6.5	50.0	50.0	52.6	47.4
Napavine	26.4	31.8	41.9	39.7	60.3	45.9	54.1
Salmon Creek	45.5	29.5	25.0	60.2	39.8	71.0	29.0
Veness						52.6	47.4
Incorporated towns							
Toledo	42.0	36.0	22.0	52.2	47.8	60.3	39.7
Winlock	44.4	43.7	11.9	59.3	40.7	56.8	43.2
Average %	36.6	29.5	32.1	39.2	56.2	54.0	46.0

[a]The total number of votes in these precincts was 1125 in 1892, 1112 in 1896, and 1020 in 1900.

riculture, rather than economic circumstances per se, seemed to be the strongest factor attracting farmers to the reform promises of the Populist party at mid-decade.

As the Populists lured voters away from the Democrats, the Republican party benefited in some of the towns. In Winlock, the Republican and Democratic presidential candidates ran neck and neck with 44 percent of the vote each in 1892, but in the elections of 1896 and 1900, McKinley raised the Republican tally to 60 percent to Bryan's 40 percent. Toledo and Little Falls voters followed a similar pattern.

As support for the reform party fell away, voting patterns became more muddled in the rural precincts. In 1900, eight precincts in the study area gave the Democrats a majority. Five of them were among the eight precincts most dominated by farming. Ferry and

Table 5. Populist vote in 1896 and population characteristics in 1900 by precinct

| Precinct/ town | % of vote for Bryan in 1896 | % of households farmer-headed | % of homes owned free | Real property[a] | | % of adults in 1900 & 1910 census |
				Average dollar value	% of adults taxed	
Salkum	78.1	95.1	80.5	473	32.7	40.3
Cinebar	77.8	94.6	73.0	408	37.8	46.6
Granite	77.4	100.0	88.9	581	38.0	48.0
Ferry	75.0	92.3	84.6	479	25.5	27.4
Alpha	67.3	97.8	89.1	555	33.1	50.0
Drews Prairie	66.6	82.6	69.6	1053	28.6	34.3
Cowlitz	61.7	93.1	80.2	1032	26.4	36.0
Prescott	60.5	76.8	80.4	638	28.3	36.8
Napavine	60.2	50.4	56.1	665	19.3	27.8
Eden	55.5	90.2	80.0	857	29.7	47.8
Little Falls	50.0	50.7	49.3	538	18.1	32.2
Toledo	47.8	20.5	49.4	290	13.0	30.0
Winlock	40.6	4.0	53.7	345	10.7	34.0
Salmon Creek	39.8	59.5	95.5	771	25.0	34.2
Ainslie	28.0	61.3	51.6	482	20.7	40.2

[a]Average value of property for those assessed taxes and identified in 1900 census; percent of total adult population taxed and identified.

Granite, however, two eastern precincts which were near the top in terms of farming, were among the top four Republican vote-getters for that year. Other factors, such as national or regional background of residents, appeared to have little relationship to the distribution of the votes in any election year.[35]

While town-country tensions are evident in the election returns, and personal politics and power struggles divided the parties locally, community residents also overcame partisanship on some issues. Virtually all Lewis county residents agreed on the need to improve farm-to-market roads and to develop a better system for road building and maintenance. Businessmen of Chehalis, Centralia, Winlock, and Toledo joined with farmers in denouncing the current state road law. A number of efforts were made on both sides to work together during the 1890s. Partisanship and conflicting values disrupted even this issue of essential unity, with farmers and businessmen developing competing good road plans and battling for control of the process. Ultimately, however, the common desire for community improvement led Lewis county residents toward compromise and working together.

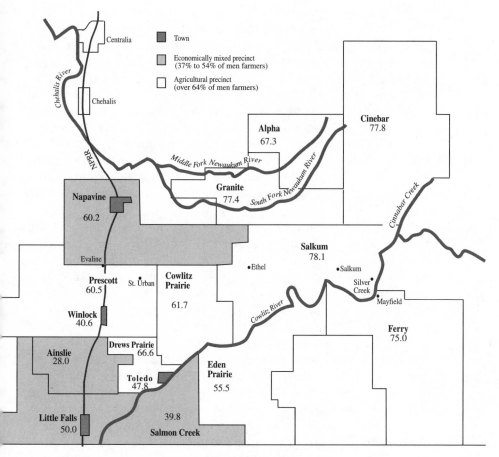

Map 3. Percent of vote for Bryan, 1896

On the last Saturday in January 1891, citizens of Toledo and farmers of the surrounding countryside held a mass meeting to discuss the road problem. They passed a resolution urging the state legislature to repeal the road law passed the previous year. That law, they complained, gave county commissioners too much power, particularly in issuing bonds without regard to county finances; it deprived district overseers of discretion in how to maintain roads and required farmers to pay their road tax in cash rather than labor, which they viewed as unjust and discriminatory.[36] Agitation over the road law continued during the next year. In early February 1892, the

County Alliance presented a road plan to the county commissioners. The initiative was generally well received. The *Chehalis Bee* praised the Alliance plan as a "sound road scheme,"[37] and Chehalis businessmen sent a delegation to the April County Alliance meeting to discuss the road question. W. A. Reynolds of Chehalis wrote a report of that meeting to which the *Bee* devoted two full columns on the front page. He said townsmen were welcomed enthusiastically as "helpers in the good cause." He went on: "I do not know what other objects the farmers alliance of the county seek to compass, but I know they are in earnest on the road question, and everybody should stand by it and encourage it to continue the agitation, the planning and working until good roads throughout our county are a reality."[38]

The good harmony did not last long. The townsmen decided to take the initiative and organized a countywide good-roads convention to be held in Chehalis. The *Bee*'s editor declared the meeting, "a representative gathering, composed of representative men from every part of the county, including a number of the heaviest taxpayers."[39] Delegates included A. J. Herren and a mix of Populists and Republicans. But the Alliance was clearly not an equal partner at the meeting, dominated by Chehalis and Centralia businessmen. The discussions at both this meeting and a follow-up held two weeks later were quite heated. In early July, Alliance leaders denounced the whole process and condemned the townsmen for turning the roads issue into one of politics and refusing to listen to the farmers.[40]

The arguments at conventions and on the front pages of local newspapers went on for months. All the agitation served to exacerbate differences and tensions between the business community and farmers loyal to the Alliance. Differences in plan details might have been worked out had not underlying political and economic differences been so great. Everyone agreed on the need for better roads to make Lewis county more prosperous and attract new settlers. The businessmen, however, saw themselves as natural leaders of the movement and by the end of 1892 must have been all too aware of the connection between the Alliance and the quickly growing Populist party that was beginning to challenge their own political dominance. They also opposed the Alliance proposal for county-owned sawmills to produce planking for the new roads. Farmers, on the other hand, viewed roads as particularly their issue. Certainly town businesses also benefited from transportation links, but farmers' prosperity depended on being able to transport their produce to market. They were unwilling to cede control of so vital an issue to others. They were also suspicious of any financing system proposed by the very bankers who stood potentially to benefit.[41]

When the state legislature passed a new bill in March 1893 that addressed many of the local concerns, the countywide efforts that had led to so much conflict faded, although good roads remained a central theme well into the twentieth century. While some portions of the laws were changed in almost every legislative session, the basic provisions of the 1893 bill remained in place for the rest of the decade.[42] In the meantime, local attention was focused on specific roads: planking (and finally graveling or paving) existing ones, building or improving bridges, grading and rerouting. In numerous efforts during the 1890s, town businessmen and farmers did work together. For instance, in 1895 and 1896 planks were laid on the road from Chehalis to Cowlitz Prairie with much of the labor and materials donated, and in 1899 a delegation of Winlock businessmen and Cowlitz farmers approached the county commissioners to urge planking of the road from Cowlitz to Ethel.[43]

Lewis county citizens belonged to national parties, voiced the slogans that their parties endorsed, debated the issues that dominated the national political scene, and voted in patterns recognizable to any student of late-nineteenth-century American politics. At the same time, many of the issues that were most compelling to people, and that were discussed in meetings and in the pages of the newspapers, were local issues. The condition of county roads, the method of tax payment, the salaries of officeholders, the personalities of local politicians, and the influence of the APA were issues that cut across party lines. They could divide members of the Farmers' Alliance as well as unite Populists and Republicans. Local party organizations also adopted their own platforms, which sometimes differed significantly from the national platforms in ways calculated to appeal to local voters. Local politics were not separate from national and state politics; they were all interconnected in ways that tended to blunt the conflicts inherent in the national party platforms.

Improving their community was something all Lewis county residents agreed on, and farmers and townspeople worked together in voluntary associations toward that end. This community structure both aided and ultimately undermined the efforts of the Farmers'

Alliance and the Populist party to lead the people in new political directions. The Alliance was patterned after the fraternal lodges that many of its members belonged to, and used the language of evangelical religion that flourished in small-town and rural America. The work ethic and gender ideology of the Alliance were consistent with the people's own, even as it pushed at the boundaries of institutionalized beliefs. It was easy for the farming families of Lewis county, as in other parts of the country, to feel at home in the Alliance and come to champion its political teachings. At the same time, no matter how committed people were to the Peoples' party, their first commitment was to their own communities. The rhetoric in Alliance halls and at political rallies may have been heated, but in the various community associations, as in their families, people practiced the art of compromise and living together.

Entire shelves in our libraries are devoted to studies of the Farmers' Alliance and the Populist party. Many books convincingly describe the intense movement culture and transforming political loyalty that climaxed in the economic and political battles of the 1890s, but few explain what happened afterward to all of that transformed consciousness and passion. A community study of this nature makes abundantly clear that after all of the studying, organizing, preaching, and voting, people continued living their lives and getting along with their neighbors. At the same time, their beliefs did not disappear. Many remained politically active. As the twentieth century brought new challenges and economic opportunities, Lewis county farmers turned to the Grange and cooperatives to work toward the goals of personal and community independence and prosperity. They also continued both to advocate far-reaching reform and to seek compromise with nearby townspeople. But as class differences hardened and war shifted popular loyalties, common ground was to become increasingly difficult to find.

ℭ𝔞 5 𝔞℣

Progressive Populists:
The Grange in Lewis County

In early 1904, D. L. Marble, an organizer for the state Grange, stopped in the country blacksmith shop of Henry McQuigg, perhaps attracted by warmth and dryness. Just a few days earlier, McQuigg had read an article on the activities of the newly formed Eden Prairie Grange. Little discussion was necessary to convince McQuigg and the farmers of his Ethel neighborhood that the Grange would benefit them as well. That tour through Lewis county was a successful one for Marble: between February 15 and March 18 he initiated six new subordinate Granges, at Ethel, Cowlitz Prairie, Silver Creek, Alpha, Evaline, and Forest, near Chehalis. By 1910, nineteen Granges were active in Lewis county, and the order continued to expand to new areas.[1] The Grange remains today an active force in Lewis county and in Washington, with a new state headquarters across the street from the state capitol campus in Olympia.

The Grange became the center of many rural neighborhoods, combining the social and overtly political in a powerful manner, similar to the Farmers' Alliance of the previous decade. In the Grange, Lewis county farmers continued to build a strong community rooted in social activities that at the same time joined their voices with those of other farmers and industrial workers, making them a leading force in Progressive era reform in the state. The Grange sought to avoid the fate of the Alliance by remaining free from close identification with any one political party, working instead with a coalition of reform forces to pass specific pieces of legislation in keeping with the general spirit of Progressivism. In a study of the Washington State Grange written in the 1930s, Harriet Ann Crawford asserts that although the Populist administration of

governor John R. Rogers had failed to pass much of its reform
agenda and the state party faded along with the national People's
party, the Populist movement in Washington did not die. Rather, in
her words, it was "buried alive and remained alive in those organizations with roots in its soil"—namely the Grange and the Farmers'
Union which was active in the wheat-growing regions of eastern
Washington.[2] Crawford subtitled her book "A Romance of Democracy." In studying Grange records, Crawford found far more than a
farmers' organization. She saw the Grange as a training ground for
civic participation that took on a life of its own and made farmers
influential in state and national events. Crawford's interpretation of
the Washington Grange is admittedly romantic, and yet it is not far
different from the conclusions of this study. The goal of a more
equitable society, the ethic of cooperation, and the spirit of participatory democracy continued very much alive in the Granges and
cooperative endeavors of Lewis county farmers in the first two decades of the twentieth century.[3]

The Grange also provided the setting to continue the expansion
of gender roles begun in the Alliance. Men and women continued
to follow a sexual division of labor that made the contributions of
both vital to the ongoing health of the neighborhood locals. At the
same time, through the Grange, rural women were able to sharpen
their political talents and join in civic life. Washington farm women
were thus important contributors to the Progressive movement, both
by nurturing a key component of grass roots support for reform and
by direct political involvement.[4]

The first Washington Granges established in the 1870s—including
several in Lewis county—apparently did not survive the depression
of that decade. A permanent Grange was organized in Washington
in 1889, spurred by widespread farmer dissatisfaction with the proposed state constitution. At its first meeting, the Washington Grange
called upon all "farmers, laboring men and taxpayers" to unite
against the proposed document, claiming it provided for extrava-
gant state spending, allowed secret legislative sessions, invited foreign investment, and would be too difficult to amend. The Grange
also endorsed establishment of a railroad commission and expressed
support for women's suffrage and prohibition, issues that remained
subjects of hot debate in state politics for the next two decades.[5]
The rapid expansion of the Farmers' Alliance with its direct link to

the Populist party stalled growth of the Grange in the state in the early 1890s, but those farmers who stuck with the Grange apparently saw the Alliance as an ally rather than as a competitor. At the 1891 state convention, an organizer for the Alliance was a guest speaker. According to the official report of the convention, the Alliance representative "gave in a very clear and forcible manner the objects of that organization, and in so doing he voiced the sentiments of the Grange. If the Grange and Alliance do not march under the same banner, they can go forth to victory side by side."[6] Two years later, the Grange sent friendly greetings to the state Alliance assembly in North Yakima. Both the number of subordinate Grange units and total membership declined steadily through the decade, and the locals that did survive were generally in shaky financial condition, but the people who remained active felt confident that when the economic and political crises diminished, farmers would turn back to the Grange. The fire insurance association that the state Grange operated was seen as a particularly valuable and attractive offering. One member expressed these beliefs in 1894, asserting: "The political excitement has hurt us. . . . But from the fact that those who have sought greater political influence by joining other orders, still have a warm feeling toward the Grange, I feel certain that when the present revolutionary excitement that is sweeping over the country takes a more rational form the grange will reassert itself and gather greater strength than ever."[7] Reassert itself the Grange did. Statewide membership nearly quadrupled in the first four years of the new century and continued to increase rapidly with the election of activist and charismatic Carey B. Kegley as state master—the order's highest office—in 1905. Lewis county contributed substantially to that growth, with most rural neighborhoods forming locals (see Map 4).[8]

Although state master Kegley had himself been a Populist, it is not clear to what extent former Alliance members participated in the creation of Grange locals in Lewis county. Of the well over one thousand Grange members whose names were found in the study area between 1904 and 1925, only six could be identified as former Alliance members. The records are far from complete, however. Most of the names come from the membership rolls of the locals that are still active. Records from the discontinued Granges and the Alliance were all lost with the passing of those organizations and the deaths of their officers.[9] Only fifty-nine Alliance members have been

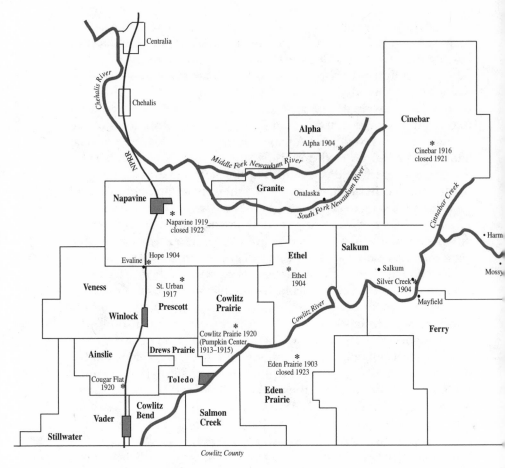

Map 4. Grange locations and year of opening

identified for the study area from newspaper accounts, although several hundred people must have participated. Therefore, many more former Alliance members than we know about might well have joined the Grange, and other sources of influence are evident. For instance, Julius Frase was elected one of the first masters of Alpha Grange, and members of his family continued active in the organization for many years. Julius was a prosperous farmer and had been a Populist party delegate from Alpha Prairie in the 1890s. It seems highly probable that he also joined the Alliance, but we have no direct evidence of that.[10] Neither the names of Judd nor Hugh Herren, who we do know were active in the Alliance, appear on Grange rolls. Sam Herren, however, another of Judd's sons, provided the first home for the Cowlitz Grange in his store in the early 1910s, and members of the extended Herren family were among the most prominent members of the reorganized Cowlitz Grange in the early 1920s. Louis Extine, the first master of Ethel Grange and a frequent officer in the State Grange during the early years of the century, also had a close connection to Judd Herren, although we do not know if Extine participated in the Alliance. Extine had made the trip from North Carolina along with the Herrens in the 1880s, and he and his wife were married on the Herren's farm in 1892.[11]

The typical Grange member in south-central Lewis county was a young, married, native-born male farmer who owned a moderately valued farm free and clear. Of those located in the 1910 census, 44 percent were native-born of native parents, but the Grange had a higher percentage of immigrants than any of the fraternal lodges, with 27 percent foreign-born, mostly from Scandinavian and German-speaking countries. The average age of these Grange members was 35, lower by several years than in the lodges or churches.[12] The Grange's emphasis on family is evident in Figure 6, but despite its acceptance of women as equal members, nearly 70 percent of Grange members in the study area were male.[13] There were more sons than daughters, reflecting the general situation in the population where men remained single and in the parental home longer than women. Husbands were also more likely to join than wives, perhaps in part because women's responsibilities for childcare made attending meetings more difficult. Married women with young children nonetheless comprised the bulk of female Grangers. At the time of the 1910 census, 60 percent of Grange women in the study area were married and under the age of 45. On average, this group had been married 12 years and given birth to four children, three of whom were still living, a situation typical of all farm women in the study area.

An ongoing controversy within the national Grange concerned what people other than farmers should be allowed as members. Country blacksmiths, teachers, and ministers had traditionally been included, but one of the charges to be leveled against the Washington Grange and its insurgent leader in 1920 was that it recruited large numbers of timber workers and other nonagricultural labor into the subordinate units.[14] Based on the evidence discussed below

Figure 6. Grange members and families posing in field near Ethel, Washington, 1910s. Photograph courtesy of Lewis County Historical Museum, Chehalis, Wash.

and displayed in Table 6, that charge was not without foundation, but was exaggerated. In the study area, 73.2 percent of the men whose names appeared on Grange membership roles during the first two decades of the century were recorded as farmers in the 1910 census; 11.6 percent of the Grange men were employed in the timber or railroad industries, but most of them were sons living on the parental farm. Less than 5 percent of household heads neither lived on farms nor engaged in a clearly acceptable profession in 1910. Possible "violators" of Grange membership rules included a railroad foreman, the landlord of a hotel with fifteen residents, and several timber workers. We have already seen, however, that Lewis county men moved frequently between farming and paid labor, so that some of these men might have been farmers at other points in their careers. Most Grange women were identified in the census as housekeepers, with about 5 percent each listed as farmers and school teachers.

Grange farmers came by and large from the middling ranks of property holders. At the time of the 1910 census, rates of clear title, mortgaging, and renting of Grange members differed little from

Table 6. Occupations of Grange members, 1910

			Occupation			
Sex	Number	% of total	Farmer	Housekeeper	Business/ professional person	Laborer
Men	251	69.5	73.2%	0%	3.2%	18.8%
Women	110	30.5	4.5	88.2	5.5	1.9

those of the whole farming population (see Table 7). Among people identified as farmers in the census, more Grangers owned both real and personal property than the norm, but the average value of Grange members' real property was somewhat lower, and the value of personal property only slightly higher than average. In this respect, Grangers differed from Alliance members identified in the 1900 census, who were wealthier than average, but the Alliance names were heavily weighted toward officers who tended to be "leading" farmers, while the Grange records held the names of all members. The average age of the Alliance members was also about ten years higher, giving them more time to acquire and improve their property. Therefore, this difference in the data cannot be taken to reflect a real variation in the membership of the two organizations. Also unlike the identified Alliance members, Grangers followed the pattern of all other association members studied in having higher than average rates of residential stability. Over half the members found in the 1910 census had also been counted in 1900.

As with the Alliance before it, a key component of the Grange's success was that it combined practical education, politics, and fun. In his report to the state Grange in 1905, Alpha Grange master Peter Griel commented that with mostly young people for members, their meetings were "not as serious and business like as should be, but what we lacked in age we make up in enthusiasm."[15] All of the locals spent a good deal of time initiating new members and practicing the ritual. Meetings usually included some kind of program which varied considerably from one time to the next, from songs and funny stories to lectures on agricultural techniques, staged debates, or discussions of political issues. The numerous other com-

Table 7. Comparison of property-holding among Grange and non-Grange farmer household heads, 1910

Grange membership	Average age	% own farm unmortgaged	Average for those assessed	
			Real property in dollars	Personal property in dollars
Grange	44	76.3	770	242
Non-Grange	49	70.7	842	218

munity associations had already schooled people in how to prepare such programs. Granges usually met twice a month. A typical arrangement was to have one meeting each month devoted primarily to business and one mostly to socializing. Food was not a feature of all gatherings, but a frequent addition, for instance if a number of new members were being installed or attendance was getting low and needed boosting. Building and fixing up meeting halls also occupied considerable attention. The halls were financed by loans and labor from members and a variety of fund-raising activities, including basket socials, bazaars, and public dances.[16]

Grange activities both cemented social ties within the organization and established a presence in the broader community. Grangers called each other brother and sister—as did Alliance, church, and lodge members. The aid the Grange gave its members reinforced the sense of family. Most locals had a sick committee which visited the ailing with baskets of food. The Silver Creek Grange in 1916 held a dance for the benefit of one member and bought a rocking chair for a sick sister. The previous year it had levied a ten cent per week fee on all members to pay for a nurse for a brother sick with typhoid. Alpha Grange donated five dollars to the repair of a brother's car in 1921, and the Cougar Flat local paid a member's doctor bill in 1923.[17] The Granges also acknowledged life transitions and holidays. Hope Grange gave a special farewell to a couple who were moving out of the area in 1913, and the Silver Creek Grange pitched in for a wedding present for two of its young members. Some of the Granges also regularly held Thanksgiving dinners and sponsored Christmas trees.[18] Ties between neighboring locals were reinforced by meeting jointly for officer installations and other special events such as plays, picnics, and dances, as well as through the

Figure 7. First Lewis County Pomona Grange, Alpha, Washington, 1910. Photograph courtesy of Lewis County Historical Museum, Chehalis, Wash.

quarterly meetings of the countywide Pomona Grange which was organized in 1910 (see Figure 7).

Grange locals were also committed to serving and improving their communities. The hall was the Grange's most visible symbol of presence in the larger community, and many locals made their meeting space available to other groups as well. Silver Creek rented its hall to the Masons, Eastern Star, Oddfellows, and Modern Woodmen lodges. Cowlitz Prairie rented its hall to the American Legion beginning in 1921 and provided it free of charge to farmers' cooperative meetings. The Hope Grange approved use of its hall for a Socialist meeting in 1912 but denied permission in the more politically charged atmosphere of 1918. Dishes, silverware, and pianos were also shared among the different organizations.[19] The Granges sometimes organized action on general neighborhood concerns such as improving certain stretches of road or telephone wiring. The St. Urban Grange looked into the matter of organizing a new voting precinct in 1919, and Hope approached the Northern Pacific Railroad Company with the suggestion of a shelter for the station at

Evaline in 1913. The Granges also provided entertainment for the general community. Many regularly held public dances. Silver Creek Grange dances were so popular that they considered building a bigger hall to accommodate the crowds in 1915. Plays were another popular, although less common, public offering.[20]

The Granges varied considerably in strength over time. Most went through periods of low involvement. Increasing attendance was a frequent topic of discussion, and the appointing of competing membership committees was a common technique to gain new recruits. Several Granges did go under completely during the early twentieth century, but many survived even the economic and organizational crises of the early 1920s. The Grange's diversity in combining social, economic, educational, and political functions appeared to be a key source of its strength. Most locals were also careful to keep their political activities strictly nonpartisan and open to a range of beliefs. Too much emphasis on any one type of activity or single political view undermined the appeal of the Grange. In the 1910s, Alpha Grange spent most of its meetings in political discussions and rarely organized social events, but attendance at meetings was quite small, rarely more than fifteen and often under ten. When the State Grange split in 1922 after the ouster of state master Bouck from the national order, some of the most active members left the Grange for the more radical Western Progressive Farmers. Thereafter, Alpha Grange grew rapidly. New members were initiated at almost every meeting in 1923 and 1924. Most of Alpha's meetings after 1922 were devoted to ritual and literary programs rather than politics, although initiative measures sponsored by the State Grange were discussed and petitions supporting them circulated. The farm families in the area apparently found political issues sprinkled into a generally social organization more appealing than an approach that focused primarily on politics.[21]

In addition to the neighborhood-based subordinate units, the Grange organized at the county level in the Pomona Grange, which met quarterly and focused on issues of larger concern. Overall, the Pomona Grange continued the emphases that had been so prominent in county politics during the 1890s. Lower taxes and a government active in the interests of producers were the two major themes in Grange resolutions. Between 1910 and 1914 the Lewis County Pomona Grange consistently passed resolutions favoring a more equitable tax structure and condemning government waste. A 1913 resolution attacked the United States court system—which routinely sided with employers against workers during this period—as a "burden" on the taxpayers that exercised unwarranted power on behalf of the "system" to the "injury of the producing masses." The state law that required road supervisors to be appointed rather than elected was frequently denounced as an usurpation of people's right to self-determination, and plans to build scenic highways while market roads were still in such deplorable condition were condemned as extravagance. The Grange supported new government action in a parcel post system, as long as it did not provide undue profits for private companies; low-interest loans to farmers; greater government control or outright ownership of railroads and telephone companies; the creation of a state-owned powder factory (to provide farmers with the powder necessary to blow tree stumps out of the ground); and better built and maintained market roads. The Pomona Grange denounced the emerging county agent system, on the other hand, as a needless expense: what farmers needed was not advice from outside "experts," but cheap money, cheap powder, and access to market.[22]

The Pomona Grange's most direct fight over taxes occurred in 1909 and 1910. In February 1909, the Lewis county commissioners declared the county's population to have grown to the point that the county was eligible for seventh-class rather than fourteenth-class status. One of the major consequences of reclassification was an increase in the allowable salaries for county officials. The Pomona Grange conducted a canvass of county population and filed suit to prevent reclassification. The battle continued until the results of the 1910 census were made known. The census revealed that, indeed, the county's population did not entitle officials to the higher salary level, and the case was dropped.[23] One Granger described the nature of the fight over taxation and county expenses in a letter to the *Bee Nugget*, echoing the old Populist theme of hard-working farmers fighting entrenched financial interests:

> Farmers all over the county are beginning to wonder why it is that the taxes keep constantly increasing, while our roads do not seem to keep pace with the tax levies, and the county getting deeper and deeper in debt. While the ordinary farmer is wondering what is the matter and where it will end, the members of the grange are getting together and doing a little figuring and thinking, with the result that many a knot hole is found

in the county fences through which the taxpayers' money is being constantly poured. One of the basic principles of the grange is the denouncing
of the credit system of doing business, and prominent members of the
grange in this county assert that now that the Pomona Grange has taken
the initiative in calling a halt upon the county administration, the investigation and education will be continued until Lewis county is free from
indebtedness and all of its business conducted upon the cash basis. This
may not prove pleasing to the financiers and moneychangers who profit
by the issue of large amounts of county warrants, but it will be a very
wholesome change for the farmer taxpayer, who, generally speaking, is
free from debt in his own personal affairs, and therefore opposed to paying a large amount of interest on a county debt in which he is directly
interested.[24]

In the decades surrounding the turn of the century, women's proper
political role was widely debated in American society. The National
American Woman Suffrage Association, the Woman's Christian Temperance Union, and many local women's clubs and associations advocated full inclusion of women in partisan politics. Most Americans,
however, continued to view politics as a male preserve. During the nineteenth century, women won the full right to vote only in Wyoming,
Utah, Colorado, and Idaho. Washington territory's legislature heard
and rejected a number of pleas to expand the franchise to women
beginning in 1854, with proponents arguing on the basis of both natural rights and the benefits that women's more moral nature could
bring to the political process and to conquering the frontier. Susan B.
Anthony toured the territory and spoke to the legislature in 1871 along
with Abigail Scott Duniway, the Oregon journalist who devoted much
of her life to the struggle for women's rights in the Northwest. After
several failed attempts, the legislature passed women's suffrage in 1883,
only to have the act overturned by the territorial supreme court in 1887
on a technicality. A new enfranchisement act passed the following year
was also overturned by the court. The all-male electorate rejected a
women's suffrage amendment to the new state constitution placed before it in 1889 and again in 1898. In 1909 the state legislature agreed
to submit the issue to voters once more at the general election in 1910.
This time the measure carried nearly two to one, passing in every
county after a generally low-key campaign—to the surprise of many.
Most other western states followed over the next four years, but women
in much of the rest of the country waited until the adoption of the

Nineteenth Amendment in 1920 to cast their first vote in state and national elections.[25]

Lewis county farm women took part in this transition, although few of them focused their political energy on the right to vote. Rather, they concentrated on class-based interests, discussing political issues and campaigning alongside men in the Alliance and Grange beginning in the 1890s. The extent to which these organizations supported women's full participation in politics varied from state to state and over time, but in Washington state both the Alliance and the Grange consistently endorsed political equality. That abstract endorsement of principle was made concrete at the community level as women in the Alliances and Granges of Lewis county took on many of the same political roles as men.[26]

In the 1860s and 1870s when the Grange was first founded, women's inclusion as full members had been highly controversial. Grange leaders defended women's prominent role with arguments permeated in domesticity, but the order continued to push at the boundaries of societal views about gender. One of the earliest historians of the Grange, James McCabe, who wrote in the 1870s, devoted much attention to the inclusion of women, defending that position against the apparently numerous critics. Women were created by God as men's equals, McCabe asserted, and they were even more in need of a respite from toil and a chance to socialize than their fathers and husbands. Furthermore, women's presence assured good conduct and the practice of moral and religious principles at the meetings. The Grange was greatly strengthened by women's contributions, and women benefited from both increased self-respect and respect from others. In the opinion of McCabe and other leaders, rural men were too apt to treat women as mere drudges, but the Grange ritual and women's participation in it emphasized their equal humanity. McCabe asserted: "In all the meetings of the Order, in all its gatherings for pleasure, the two sexes are brought together, and placed upon an equality, and the farmer is thus quietly and forcibly reminded that his wife and daughters are ladies entitled to all the courtesies and attentions of polite society, and not mere drudges charged with the performance of household work; something he has been too apt to forget."[27]

The kind of equality that McCabe and others pronounced accepted the belief that male and female attributes differed naturally, even though men and women shared a common humanity. Women provided a morally healthy tone for the meetings and proper attention to education for the young. And, McCabe took pains to assure his readers, a Granger's respect for women did not translate into endorsement of the woman's rights movement. Women were far too refined for activities such as voting. He quoted at length a Captain E. L. Hovey, who said:

> Some who are inclined to see a humbug in every new move assert that this is a "woman's rights" movement; others that it is a cover for political intrigues. Nothing could be further from the truth. The fact that women are admitted to full membership in the Order I regard as one of its most worthy features. I do not believe in woman suffrage, nor never can. I do not believe in making a plow-point of a gold watch; but the condition of a people, its customs, its manners, its morals, its social standing, its educational status, depend more upon its women than upon man. Is there not as wide a field for improvement in woman's sphere as in man's? Besides, when men are assembled for mental culture or social chat, what more stimulates them to high-minded action than the presence of woman?[28]

Lest anyone still fear, McCabe assured his readers that women in the Grange were no more likely to take up woman's rights than women at their own firesides, "for at each place she is in company with her husband and brother."[29] Not all Grangers, however, agreed with the limits McCabe set on women's activities. Donald B. Marti, in a study of the late-nineteenth-century Grange, found a substantial minority of members did advocate women's suffrage, generating substantial debate. Between 1885 and 1900, the National Grange endorsed and then withdrew endorsement of women's suffrage two times, before finally giving permanent approval in 1915.[30]

The Washington State Grange was far more liberal than the National on most issues, including women's suffrage. It endorsed suffrage from its formation and regularly reasserted that support. Women were frequent speakers on a variety of issues at all state conventions from the beginning. A range of beliefs on women's place in society was expressed, but even those with a fairly traditional view of women's role advocated expanding their influence. In 1891, Sister M. Still demanded the right of franchise so that women might "crush this hydra-headed foe," liquor. In 1899, the Education Committee submitted a resolution, which was approved by the larger body, stating that since women were the natural educators of children, more women should be elected school directors. Laura Fitzgerald echoed what was increasingly becoming the dominant argument of the National American Woman Suffrage Association when she proclaimed that her reluctance to "meddle in politics" had been overcome by the belief that if mothers voted, a far better world would result. She stated:

> I have never been a very strenuous advocate of woman meddling in politics, for I thought her burdens were more than she could do justice to, even if she were well educated for it. And the reason I would now have mothers vote, is because, if she did, the babies would be represented in congress. And a congress of mothers would not vote for thirty years gold bonds to enslave their children. Women would abolish child labor; and pension orphans. We would have healthy school rooms and play grounds. We would protest against military organizations in school, which lead children to think that war is not a shame and a national disgrace.[31]

Washington Grange women also argued a woman's rights position from the standpoint of common intelligence and humanity, side-stepping the issue of male and female difference. Mrs. H. E. Wing proclaimed women's chief work in the Grange to be political—to rouse the farmer to see the injustice of his overtaxed position and overthrow the old parties which were controlled by monopolies and trusts. Mrs. Rima, in a pointed departure from the usual definition of "independence," urged in 1894 that boys as well as girls be taught to bake bread, cook potatoes, and mend clothing so that they might lead independent lives.[32] Jennie Jewett submitted numerous resolutions over the years, including one not to back any legislative candidate who would not work for women's suffrage, and one to amend state law so that women would have the same property rights at the death of a spouse as men. On the issue of the franchise she spoke frequently, in one address stating, "If there is any reason why men should be permitted to exercise the right of suffrage, the same reason exists why women should be permitted to do so. Sex has no more to do with suffrage than it has to do with going on a journey or going to church."[33] Mrs. C. A. Buchanan urged Grange women to join the fight against the injustices farmers faced in 1891, saying: "Sisters, we may not be as good in the art of driving nails with steel hammers, such as our brothers use, but with the hammer of intelligence and right we can drive the nail of justice and right and fair dealing into the coffin of political infamy and corruption—the coffin of dismantled ships of capitalist combinations which have been sailing over our rights for these many years, and having done this we will, by our united strength, bury it forever from our sight.''[34] The State Grange participated in the broad-based effort by Washington women to gain the vote in 1910. Thereafter, state master Carey B. Kegley continued to fight at the national level, chairing the standing committee on women's suffrage in the National Grange during the 1910s and urging the order to endorse publicly a federal constitutional amendment.[35]

Despite the emphasis on human equality, Grange locals practiced a gendered division of labor, with women specializing in preparing food, coordinating sick relief, and providing musical and literary entertainment, while men specialized in financial oversight and the public leadership represented by the Grange master. Such a division was clear in the Silver Creek Grange in the spring of 1907 when a committee of three women was appointed to purchase new officers' sashes and a committee of three men to write new bylaws. Several years later when Silver Creek built a new hall, the building committees were composed of men, and the women met after construction to wash windows, black the stove, and "put the hall in proper shape." When the Hope Grange was asked to provide food for a nearby Farmers' School in 1915, "it was left to the Sisters to talk it over after the meeting." Although in this case discussion of "women's work" was not deemed worthy of general business time, the importance of women's work was affirmed by Grange members in other instances. Clara Layton of Eden Prairie Grange reported to the state assembly in 1907 that "the women folks generally take the lead" in raising money through selling cushions and organizing ice cream parties, box socials, and dances. She added, "We think were it not for the women of Eden Prairie Grange the men could not do so well."[36]

Public representation of the Grange, on the other hand, was considered men's responsibility. Although in a typical year three or four locals across the state had women masters,[37] the highest office in Lewis county Granges was always filled by a man in the first two decades of the century, even in locals that had especially strong and vocal women. Emma Uden served as lecturer and secretary in the Alpha Grange during the 1910s, and was clearly a leader of some vigor. In 1923 she was elected president of a three-county district of the Western Progressive Farmers representing Lewis, Grays Harbor, and Thurston counties. She was never elected master in the Grange, however, perhaps in part because her radical views were feared to be too divisive, but as lecturer she influenced discussions even more directly. Her sex was clearly a major factor.[38]

Men also took charge of the cooperative endeavors undertaken by many Grange locals. Cooperative buying was discussed in the general meetings with women present and sometimes participating, but purchasing committee members and agents were always men. Most of these cooperative efforts were small-scale, without permanent organization. A group agreed to pool money to buy some commodity in bulk and appointed an individual or committee to make the purchase. For example, the Alpha Grange bought clover seed, sugar, and flour in this manner in the early 1910s, buying flour jointly with Silver Creek on at least one occasion. St. Urban made a series of hay purchases in 1918 and 1919. Members were so pleased with the quality and the savings they achieved that they appointed a three-man committee to look into the possibilities of building a cooperative store. The Hope Grange actually constructed a grain storage warehouse in 1914, although what became of it is unclear. The trustees were empowered to rent it out at their discretion in 1915.[39] This division of labor probably reflects the division that took place on family farms. If the farm husband generally handled major purchases for the family, few apparently questioned that the men should handle similar purchases for the group. Finance, by-laws, and building committees were usually all male as well in most of the Granges.

Much of the central work of the Grange, however, including political activism, education, and community building, was considered to be the work of both sexes. In this respect, the Grange differed significantly from the lodges, which were organized to ensure that women remained in limited and subordinate roles. Men more often made and seconded motions during Grange business meetings, but they also constituted a majority of members. Women's smaller numbers apparently did not intimidate them, for they took part in discussions of almost all issues. Hattie Antrim was among the most active members of Hope Grange. She was fifty-six years old when she and her husband Peter joined Hope Grange in its founding year, 1904. They had earlier subscribed to the *People's Advocate* and later supported the Socialist party. She gave birth to nine children, several of whom continued to live on the family farm as adults at the time of the 1910 census. She made or seconded motions over the years on such issues as taxes, the work of county commissioners, and the Rural Credit Commission, and served on a number of committees. At one meeting in 1912 when the lecturer was absent, she took the chair and led a discussion of the initiative and referendum measures then before state voters. In 1910 she was elected an officer in the Lewis County Pomona Grange.[40] Addie Huntting and Mattie Tucker of Silver Creek and Emma Uden of Alpha were similarly involved in a broad range of issues in their local Granges. Uden even served on a good roads committee in 1920—an appointment that would have been highly unlikely in the 1890s when roads were a distinctively male issue.[41]

The lecturer occupied a particularly influential post in the Grange, being responsible for member education and determining in large part the content of the program after routine business was dispensed with. The Silver Creek, Alpha, Eden Prairie, Hope, and St. Urban Granges all had women lecturers during the early twentieth century.[42] Women also chaired committees with male members, and frequently served as secretaries, treasurers, and chaplains. Absent officers were often temporarily replaced by a member of the opposite sex, suggesting that in most cases positions were not viewed as having gendered attributes. In the State Grange, women frequently served as secretary and chaplain during these decades, and usually held one of five seats on the executive committee. Of the standing committees, three-quarters had at least one woman member during the 1890s, but that figure fell to about half in the early twentieth century as membership mushroomed.[43]

The traditional assignment of duties on the basis of sex was also sometimes challenged in Lewis county locals. Relief committees often had at least one man, though the majority of members were usually women. Food preparation became a sensitive issue in some Granges, although there are many recorded instances of women preparing and serving food without question. Silver Creek sisters made lunch for the brothers hauling gravel for road work in 1920, but when a similar division was suggested for a work party in the St. Urban Grange that same year, a male member proposed that the women don overalls and join in the road building instead.[44] In 1922, Cougar Flat women suggested that the men alternate with the women in preparing dinner, and at a later meeting proposed that the men prepare the food baskets for two ailing brothers, clearly challenging the usual division of labor. Ethel Grange minutes do not record any discussion of the issue, but all-male committees served at least two suppers in 1919.[45]

The St. Urban and Cougar Flat Granges had women's auxiliaries in addition to the mixed-sex meetings. Mrs. H. W. Paschke suggested creating the St. Urban auxiliary in 1920 "so more of the Ladies with children could be present." Although children did come to Grange meetings and teenagers joined the order from the age of fourteen on, attending an evening meeting with young children could be difficult. The auxiliaries held informal home meetings during the day. The St. Urban auxiliary made craft items and held bazaars to raise money for general Grange needs, and in 1921 organized a drill team. The Cougar Flat auxiliary also raised money, in 1922 paying the master's way to the state convention.[46] The Grange auxiliaries differed significantly from the auxiliaries of the lodges. In the Masons and Oddfellows, women were not allowed at all in the primary group, and both men and women participated in the Eastern Star and Rebekahs. The Grange, on the other hand, did not ban women from the general meeting with the creation of auxiliaries. The auxiliaries did not help maintain a male monopoly on power in the order; rather, they provided women with young children a more convenient way to contribute.

Through the Grange, rural women were able to gain a voice in the largely male world of electoral and legislative politics. Other women in other times and places had also gained such access through activism in the abolitionist movement, the prohibition movement, suffragist organizations, labor unions, and urban social reform groups.[47] Despite these challenges, male domination of public life generally and of the political parties in particular remained secure throughout the nation. Without the Grange and the Farmers' Alliance before it to accept and nurture women as political actors, it is doubtful that the rural women of Lewis county would have had much political voice. The articulate oratory of women in the political struggles of the 1890s and the struggle for suffrage put to rest any lingering doubts in local farmers' associations about women's political competence. Political issues carried no gender assignment in the Grange. Both men and women were expected to be involved. Grange women in Lewis county participated in discussions of political issues, made and seconded motions on a variety of topics, circulated petitions on initiative measures, and served as officers and delegates. Alliance men had been proud of the political expertise of women and had frequently mentioned in their newspaper reports that Alliance women were even more concerned than men with "the issues of the day." To them, this had been a surprising innovation. In Lewis county, however, Grange records in no way make an issue of women's political involvement. In these twentieth-century Granges, women's participation in politics had come to be taken for granted. Moreover, women took part in politics as members of the farming community. Their interest was not limited to "women's issues," but was in the same range of class- and region-based issues that men addressed.

Perhaps the sexual integration of politics helps explain the low-key nature of the 1910 suffrage campaign in Lewis county. The county as a whole and most Grange communities within it endorsed the suffrage amendment to the state constitution, but the issue does not appear to have generated much excitement locally. State master Kegley urged "that every Subordinate Grange . . . be a strong equal suffrage association," and the 1910 state convention adopted six resolutions on equal suffrage. Of the subordinate Grange minutes examined for this study, however, only the Silver Creek records included a formal endorsement of suffrage, passed unanimously in 1908.[48] Emma Smith DeVoe, the president of the Washington Equal Suffrage Association, spoke before the Lewis County Pomona Grange in May 1910, but the newspaper accounts did not record any action taken as a result, although they did report several resolutions on other topics. The Eden Prairie Grange debated at its next meeting whether women's suffrage was an appropriate topic for the Grange to consider. Some members saw it as a purely political question and therefore out of line, while others disagreed. The following month at the State Grange meeting, the master of the Lewis County Pomona, C. W. Frase of Alpha, introduced a resolution condemning delegates to the National Grange who opposed equal suffrage.[49] Otherwise, little mention was made of the issue in Lewis county records. The county press was virtually silent on the issue leading up to election day, and the county Republican and Democratic platforms, as reported by the *Bee Nugget*, made no mention of women's suffrage.[50] Either opinion was too evenly divided for politicians and editors to risk taking a stance, or the issue had been around too long to arouse much interest. Consolidation of schools, the initiative and referendum, and the Pomona Grange injunction against county commissioners who had tried to raise their salaries were all much hotter topics at Grange meetings and in the county press during the 1910 campaign season.[51]

Eleven of nineteen precincts in the study area returned majorities in favor of granting women the franchise (see Table 8). While not an overwhelming endorsement, it was a tremendous advance over the 1898 vote when only Cinebar and Granite among the study

Table 8. Percentage of 1910 vote for suffrage, 1914 vote for prohibition, and nativity of adults

Precinct/town	1910 for woman's suffrage	1914 for prohibition	Native-born parents	Germanic immigrant[a]	Scandinavian immigrant[b]
Precincts supporting woman's suffrage					
Ainslie	93.3	69.1	37.7	1.2	42.0
Salkum	76.2	66.7	62.7	4.5	5.7
Stillwater	75.0	64.4	38.5	6.2	10.4
Ferry	75.0	46.4	72.0	5.3	9.3
Alpha	67.9	53.0	32.0	11.8	13.1
Cinebar	62.5	39.1	43.4	9.4	8.5
Winlock	60.5	61.1	50.4	9.3	6.8
Little Falls	58.8	49.5	45.7	3.0	4.5
Drews Prairie	57.1	50.0	50.0	6.6	3.9
Ethel	56.0	61.8	53.4	2.5	18.6
Eden	53.8	63.0	67.0	5.2	7.8
Precincts opposing woman's suffrage					
Granite	46.7	62.0	68.4	3.5	8.8
Cowlitz Bend	44.4	65.6	NA	NA	NA
Veness	44.4	54.7	23.9	15.9	41.5
Toledo	41.0	66.0	77.2	2.0	1.6
Salmon Creek	40.0	59.4	71.2	3.1	0.4
Napavine	37.2	67.5	49.5	10.7	10.2
Prescott	29.4	36.9	28.0	28.0	19.3
Cowlitz	14.3	33.3	50.6	13.2	5.5

[a]Includes German, Swiss, and Austrian.
[b]Includes Norwegian, Swedish, Danish, and Finn.

precincts had favored women's suffrage. The lack of publicity on the issue in the county press was reflected in the vote totals. Less than half as many men voted on the suffrage issue as voted in the congressional race on the same ballot. Grange membership apparently was a factor in the prosuffrage vote, as would be expected. The seven precincts whose residents attended Silver Creek, Alpha, Ethel, and Eden Prairie Granges endorsed women's enfranchisement, with the exception of Granite, which split nearly evenly on the issue. In contrast, the three precincts with residents attending Hope Grange —Napavine, Veness, and Prescott—opposed suffrage. The large German-speaking population in this area might well have been a factor. Neighboring Cowlitz Prairie, where the Grange was not permanently organized until 1920, also had a large German-speaking population and gave the least support of any of the study precincts to suffrage. Studies of midwestern communities have shown strong opposition to women's suffrage among German immigrants, whether Catholic or Lutheran, because of religiously based gender beliefs and fear of prohibition.[52] The German Lutheran church located just outside Winlock, which drew its members from Prescott and Veness German immigrant farmers, emphasized its belief in gender differences by having men and women sit on opposite sides of the sanctuary. Both Prescott and Cowlitz precincts also housed Catholic churches which served the German-speaking residents. Customs of beer brewing and drinking were also well established locally. The desire not to offend the large and well-regarded German population might also have been a factor in the quietness of the campaign.

Not all Germans opposed women's suffrage, however. Populists Charles and Anna Wuestney who had briefly edited the *People's Advocate* during the 1890s were German immigrants who strongly endorsed women's right to vote. And Pomona master C. W. Frase, who advocated suffrage at the State Grange meeting in 1910, was a member of an extended German immigrant family prominent on Alpha Prairie. Alpha was the only precinct in the study area with German-speakers making up over ten percent of the population to support women's suffrage. It was also the farthest removed from the immigrant churches clustered near Winlock and Toledo that helped maintain a distinctive German identity.[53]

Outside of the German population, opinion on prohibition was generally not a factor in the equal suffrage vote, despite the general supposition that women voters were more likely to favor restrictions on alcohol. In April 1910, the men of Little Falls had voted down a local prohibition measure, but that November they supported women's enfranchisement. Citizens across the state were given the chance to vote on prohibition in 1914. Of the five precincts that voted against prohibition, three had supported women's suffrage in 1910, while six precincts that favored prohibition had opposed suffrage (see Table 8). Clearly in Lewis county, men did not base their decision on women's suffrage primarily on their feelings about liquor.[54]

Party affiliation also had little to do with support of equal suffrage. Of the four most heavily Socialist precincts, two favored and two opposed the amendment. Four precincts that gave majorities to the Republican congressional candidate favored suffrage, and five opposed. The two precincts that gave majorities to the Democratic candidate were split on the suffrage issue. County party leaders had apparently been prudent to ignore the equal suffrage amendment.

The response of the political establishment to the enfranchisement of women was ambivalent, oscillating between matter-of-fact acceptance and condescension. In 1912, when the Republicans nominated Mrs. Bertha Gage for county clerk, the *Bee Nugget* described her as "a most capable woman." The voters agreed. She won an easy majority, running well ahead of most of the rest of the Republican ticket. Chehalis Republican women organized a Taft club that same year, at the urging of male party leaders. The Bull Moose convention of 1912 included eight women out of one hundred delegates.[55] When two hundred women in eastern Lewis county organized a Women's Good Road Association, the *Bee Nugget* responded, "Suffragettes all over the world will be interested in this practical result of women's political enfranchisement and all members of good roads associations will be deeply in sympathy. . . . Their example should be and doubtless will be, followed by their sisters in other localities."[56]

On the other hand the *Bee Nugget*'s editor continued as in the past to deride the political competition in part by implying that its support came from women. In reporting on a speech made by Stanton Warburton, the area congressman elected as a Republican, but since bolted to the Bull Moose party, the newspaper claimed that only fifty-nine people came out to hear him, most of them curious Democrats and regular Republicans, and twelve of them merely women. In September 1911, before the first election in which Lewis county women could vote, the *Bee Nugget* exhibited a decided lack of respect for the competence of women voters in an editorial urging women to vote according to their own minds, rather than following the dictates of their husbands and sweethearts.[57]

Women's suffrage was only one of the issues falling under the rubric of Progressive reform in which the Washington State Grange was heavily involved. The Grange retained a carefully nonpartisan stance, its members remembering full well the disintegration of the Alliance, and the rhetoric of the period was certainly tame compared to the fiery orations of the Populists. Nonetheless, Washington state farmers had not become shy of politics after the battles of the 1890s. The crusading reformer Kegley was reelected state master by large margins until his death in 1917, and he was followed for several terms by the avowedly radical William Bouck. At the state and national levels, Washington Grange leaders were at the fore-front of reform politics. The urban middle-class aspect of Progressive reform is the side of this complex and difficult-to-categorize move-ment that is probably best understood by historians, but the heart of the Progressive movement in Washington was the alliance among organized farmers, organized labor, and their urban middle-class supporters who sought a redistribution of the benefits of an indus-trialized economy.[58]

Together with the State Federation of Labor, the Farmers' Union, and other organizations, the Grange successfully lobbied for an im-pressive array of legislation, including the direct primary and an advisory primary for senator in 1907; a direct legislation amendment to the state constitution, the eight-hour day for women workers, the beginnings of an industrial compensation system, a state pure food and drug act, and ratification of the federal income tax amendment in 1909; the women's suffrage amendment that was ratified in 1910; and the referendum and recall in 1912.[59] In 1909, one of their most successful years, the Grange, Farmers' Union, State Federation of Labor, and Direct Legislation League (a Seattle-based group) or-ganized the Joint Legislative Committee (JLC) in an effort to make their reform coalition permanent. Robert Saltvig, in a study of Pro-gressivism in the state, called the JLC "effectively a third house" in the legislature.[60]

This reform drive was successful in part because state leaders with quite different political philosophies were willing to work together on specific measures. A number of studies of state politics have iden-tified the factions and shifting alliances within the groupings iden-tified as socialists, labor leaders, agrarians, and urban reformers.[61] The Progressive alliance included radical, moderate, and relatively conservative wings in each of these groups which fought behind the scenes for control. Conflicting interests and suspicions were also abundant between the arid and largely rural east and the heavily urbanized west side of the state. These struggles as well as political circumstances led the JLC to follow different strategies at different times. Throughout the 1910s it remained for the most part non-partisan, identifying progressive candidates in the primaries of the two major parties and uniting behind a single candidate who would accept their reform package for the general elections. After having temporarily lost ground to the Populists, the Republican party regained its hold on state politics after 1900. Therefore, much of the reform legislation was accomplished through the help of Republican "insurgents" who shared with the producers' alliance a desire to overthrow the old guard.

With the passage of the direct legislation amendment, the JLC turned more attention to direct voter appeals. It launched an ambitious initiative drive, popularly known as the Seven Sisters, in 1914. Five of the seven proposed initiatives appeared on the state ballot, two having failed to gain enough signatures in the petition phase. The initiatives reflected the long-held agrarian concerns with taxes and efficiency in government, as well as specific labor issues. The proposed laws would have inhibited the sale of fraudulent stock; abolished the state bureau of inspection; prohibited employment agencies from charging laborers fees; added first-aid provisions at employers' expense to the industrial compensation law; and lowered the state road construction levy. Two additional initiatives not backed by the JLC were also on the ballot that year: state-wide prohibition, which was officially sponsored by the Anti-Saloon League and also supported by many Grangers, but opposed by most labor groups; and a universal eight-hour day sponsored by the Socialist party and widely backed by labor unions. A well-financed opposition campaign calling itself the Stop-Look-Listen League generated enough negative publicity to largely defeat the initiatives. Only the employment agency and prohibition measures passed.[62]

In 1912, state voters gave a plurality of votes to Theodore Roosevelt, who had broken ranks with the Republicans and formed the Progressive party. Roosevelt's candidacy was backed by many reformers nationwide, including Kegley and the man who was to be his successor as state master, William Bouck. The Progressive gubernatorial candidate in Washington became involved in a messy divorce scandal and was unable to carry the state. The governor's seat went instead to a moderate Democrat, Ernest Lister, who had been a member of the Rogers' Populist administration. Lister had not, however, retained a Populist fervor for reform. In the midterm elections of 1914, the same year the JLC's Seven Sisters initiatives were largely defeated, conservative Republicans regained control of the legislature and, unhindered by the governor, repealed much of the reform legislation that had been passed earlier. Two of the laws passed in 1915 required initiative petitions to be signed in the presence of a registration official and required voters to register every four years, measures widely regarded as attacks on direct legislation and rural voters who had never before been required to register. Other new laws prohibited picketing during labor disputes, restored the caucus primary (which was believed to give the advantage to party insiders), and allowed only tax-paying property owners to vote on bond issues. The JLC filed referendum petitions on many of the new measures, which went before the voters in 1916. This time the JLC's campaign was far more effective. Laws that would have seriously undermined direct legislation and recall and that banned picketing, as well as several others, were overturned by big majorities.[63]

Granges in Lewis county generally supported the state legislation and initiatives endorsed by Kegley and the Joint Legislative Committee, but only after their own thorough discussions. They listened to talks from their own lecturers and visiting speakers, read letters from numerous organizations, and held lively debates on issues. Members of Hope Grange, most of them property owners and tax-payers, hotly debated whether nontaxpayers should be allowed to vote in bond elections before the 1916 referendum vote. After a long discussion, they reached no agreement and laid the matter over to a future meeting.[64] Most Grange locals in the study area endorsed the 1912 ballot measures that legalized the initiative, referendum and recall, and the Seven Sisters initiatives that followed in 1914.[65] The county's leading newspaper, the Bee Nugget, meanwhile persisted in condemning direct legislation, despite vociferous opposition to its editorial stance by readers and the over two-to-one majority that Lewis county voters gave to direct legislation in 1912.[66]

Lewis county voters on the whole followed the general state pattern, with only seven of the now twenty precincts in the study area supporting the JLC-backed initiatives in 1914.[67] They included the five precincts with the highest percentage of Grange members: Eden Prairie, Alpha, Ethel, Granite, and Salkum. Ainslie and Veness precincts also favored the JLC initiatives. These were economically mixed areas that contained relatively few Grange members but strongly supported the Socialist party. These latter two were also the only precincts to give a majority to the eight-hour day initiative, despite the large numbers of timber workers in other precincts and the popularity that the eight-hour day movement was to gain among loggers during the coming war. Prohibition won big, with only six precincts failing to pass it, three of them in very tight races.

Despite the strong support for prohibition in the county, it remained the highly divisive issue it had been since the 1890s. Many respected farmers and businessmen, particularly those with immigrant backgrounds, held traditions of moderate drinking they saw no reason to discontinue. Local chapters of the Woman's Christian Temperance Union (WCTU) and similar organizations, on the other hand, continued to rally and distribute propaganda on the unmitigated evil of liquor. The Centralia saloon patrol, a mixed-sex group, launched an assault on the town's saloons in 1901. An angry exchange of letters in the papers ensued. Mrs. M. Reynolds wrote in one of those: "Every true mother knows and dreads [a] flame fiercer and more deadly than any flame of physical torture and she knows too that the licensed saloon is the hot bed which nourishes that flame. Shame on the pseudo modesty which prevents so many of our good women from coming out and declaring war to the death over this monster evil, because, forsooth, some men or beings having the form of men . . . are ready to point their fingers and cry crank!"[68]

With the passage of a local option bill in the state legislature, advocates of prohibition launched campaigns in a number of individual towns to close down the saloons. Between 1910 and 1913 prohibition votes were held in Centralia, Chehalis, Dryad, Pe Ell, Little Falls, and Napavine, with the "wets" winning in every case but Pe Ell. The move to incorporate Napavine in 1913 was largely a battle of wets versus drys, as reported by the *Bee Nugget*, with the wets winning.[69] One of the issues was the perceived danger posed by large numbers of single male mill and logging-camp workers gaining access to alcohol. In December 1912, the county commissioners denied a license to a saloon in the logging town of Dryad after receiving a petition signed by businessmen, pastors, the school principal, and the mill owner. The commissioners declared their reluctance to allow saloons in towns where the working men depended largely on a mill.[70] The Chehalis chapter of the WCTU held religious services and distributed "comfort bags" to single men at logging camps—assuming that the married men were looked after by their wives. The bags contained sewing and first-aid supplies and a Bible as armor against the temptation of drink.[71]

Drinking was also an issue in the Granges and lodges. The Silver Creek Grange suspended a member for engaging in the saloon business in 1907, and in 1920 finally discontinued its public dances after repeated troubles with drinking.[72] The Toledo Masons twice tried a member who was a bartender for selling alcohol, acquitting him of "unmason-like conduct" in the first trial, but suspending him the second time in 1908. The Cougar Flat Grange, on the other hand, rejected a motion in 1923 to investigate a brother for selling liquor.[73]

Overall, the political rhetoric of the prewar years was tame. Grange members considered themselves members of the producing classes and were wary of proposals that might benefit the "moneyed" classes at their expense, but there was not a strong sense of oppositional politics. Reading Grange minutes and accounts of meetings in newspapers gives the impression of sober discussions of practical issues, rather than the kinds of fiery speeches and debates that enlivened Alliance meetings in the 1890s. Subordinate Granges supported the reform initiatives of Kegley and the State Grange, but meeting minutes rarely record any mention of the alliance with labor or conflicts with townspeople. That solidarity and conflict was more evident at the Pomona level. For instance, in 1914 the Pomona passed a resolution condemning the massacre of women and children in a Colorado labor dispute. At the August 1913 meeting, Grangers discussed at length their problems with merchants, particularly in Centralia and Chehalis. One proposed resolution condemned merchants for sending outside for produce, rather than marketing locally grown products. Many members reported "bad treatment" by merchants, although specific problems were not recorded. Ill-feeling was high enough that both a boycott and the building of a Grange store were discussed as possible remedies, but not acted on.[74] Despite such evidence, the clear articulation of issues that was so apparent in the Farmers' Alliances in the 1890s was generally lacking in the Granges during this period. Resolutions were scattered in focus, rather than organized around a coherent philosophy.

The lack of a coherent philosophy reflected the generally eclectic nature of the Progressive reform movement and was due in part to reformers being scattered in several different political parties. Both the Democratic and Republican parties contained strong Progressive wings. Theodore Roosevelt himself was generally identified as a Progressive and led the reform wing of the Republicans into the Progressive, or Bull Moose, party in 1912. He had the support of such diverse reformers as Jane Addams and the Washington State Grange's Carey B. Kegley, although the widely read Socialist newspaper, *Appeal to Reason*, branded him a "political trickster and a ranting demagogue" as well as a "tool of the trusts and foe of the people."[75] The Democrat's national platform continued to show Populist influences in its support of popular democracy and economic justice, despite the party's inclusion of a racist southern elite. It denounced imperialism and the suppression of national liberation movements abroad, and advocated a decrease or end to the protective tariff that indirectly subsidized American industry, a progressive income tax, regulation of large corporations, and increased support and protection for workers and farmers.[76]

The Socialist party was perhaps the vehicle most likely to articulate a position of radical protest in the early 1910s and to inherit the Populist vote.[77] Organized in 1901 and headed by labor leader Eugene Debs, it attracted many former Populists in the Midwest, Southwest, and West. It combined the popular American emphases on democracy and individual liberties with Marxist theory, differing markedly from the more doctrinaire and urban immigrant-dominated Socialist Labor party. The Socialist party advocated an electoral route to a reordered national economy in which the large-scale means of production and distribution would be collectively owned and democratically controlled. It also supported full political and social rights for women, and a number of reforms often included in the Progressive agenda such as the initiative, referendum, and recall, and worker protection. The Socialist party's loose organization allowed for local variations, and many of its leaders emphasized immediate reforms rather than long-term revolution, making Socialism attractive to many reformers, while at the same time blurring the distinctions between Socialists and more moderate Progressives.[78]

Milwaukee and other midwestern cities with vibrant labor movements were strongholds of the Socialist party. It also had a large rural wing, particularly among Oklahoma tenant farmers. In Lewis county, Socialist candidates ran well in a number of the farming and mixed timber-farming precincts in the study area between 1908 and 1916, but the Socialists lacked the widespread social and cultural base that had given the Alliance and People's party such power and made the state Grange so effective. James Green and Garin Burbank, in their studies of southwestern rural Socialists, make a strong case for a well-developed consciousness of class among Socialist voters in Oklahoma, and for the use Socialist organizers made of existing cultural forms in their appeals to voters.[79] The agricultural economy of Lewis county was quite different, however, without a large body of tenant farmers sharply distinguished from landowners, and the limited evidence available does not indicate a strong grassroots Socialist organization in the county.

The Socialist party in Lewis county was able to run full county slates in 1908 and 1910 and sponsored a series of lectures by national speakers in Chehalis in the winter of 1912. The staunchly Republican *Bee Nugget* was surprisingly moderate in its reports on the series, deeming the first lecture a "success" and mildly praising the last speaker for sticking to the issues at hand. Not all reaction was so tolerant, however. A Socialist speaker touring eastern Lewis county the following summer was interrupted in the middle of one talk when the foreman of the local lumber mill kicked the soap box on which he was standing out from under him. In the ensuing fistfight, the Socialist, a former professional boxer, was the clear victor, and he resumed his speech to the hearty cheers of the assembled audience.[80]

Although timber workers might have seemed its natural base, most of the Socialist party's support in Lewis county apparently came from farmers. The *Bee Nugget* in a 1911 article declared the philosophy doomed because "conceit and human selfishness" are inborn qualities but said Socialism was nonetheless popular in the rural districts. Another article in the same edition attributed a resolution passed by a local Grange opposing military involvement in Mexico to "the activity of some of its members with socialistic ideas."[81] Some Grange members actively promoted the Socialist party. In Hope Grange, Hattie and Peter Antrim sponsored several resolutions in support of socialist organization. Their Grange had some correspondence with a Centralia Socialist local, and allowed use of its hall for Socialist meetings. Jessie Swope farmed with her husband and two small children on Eden Prairie, and was the sister of Eden Prairie Grange master Thomas Larter. She not only subscribed to the national Socialist newspaper, *Appeal to Reason*, but in 1913 paid for 200 copies of the paper to distribute locally. For the most part, however, the Granges stayed true to their nonpartisan position, and steered clear of any direct involvement with the party.[82]

Letters to the editor were apparently one of the main means of voter education on Socialist principles in Lewis county. Between 1911 and 1913 letters explaining Socialist ideas were so frequent and lengthy that the *Bee Nugget* finally declared a word limit.[83] One of the frequent correspondents was Carl Motter, a prosperous farmer near Chehalis who had grown up on Cowlitz Prairie where his father, David, had been a member of the Farmers' Alliance and People's party, as well as a Methodist sunday school teacher. His wife, formerly Minnie Griffin, had also been active in the Alliance. Will Hopkinson, county secretary of the party, also wrote in frequently. Both Motter and Hopkinson emphasized inequalities in the distribution of wealth and the small return producers made on their labor. The remedy, they claimed, was "the collective ownership and democratic management of all publicly used tools of production and distribution." They frequently cited the successes of the Milwaukee Socialist administration as proof of socialism's potential.[84] They did not convince all their readers, however. Judd C. Bush, who was to become one of the early organizers of the Lewis-Pacific Dairymen's Association, claimed that farmers supported the Socialist party only because they did not really understand its teachings, and since they believed first and foremost in their homes and country, they would renounce it once its true nature was made known. Bush wrote:

> I am yet to be convinced that there is merit in the idea of collective ownership and democratic management of all wealth. There are scattered among the farmers many who think they are socialists and as long as socialism is as elastic as it appears to be at present they can easily pose as such, but when it comes to the test, if the doctrines preached by many as pure socialism are the real tenets of this closely woven secret political society, there will be a great falling off from the socialist column of the toiling agricultural masses who believe in home and country first and last.[85]

Motter replied the following week, claiming Bush was really opposed to democracy, and so was trying to subvert the one true advocate of democracy, the Socialist party. He also emphasized the difference between publicly used tools of production and articles that rightfully should be privately owned:

> Now I don't blame Mr. Bush for being against the collective ownership of all wealth, as tooth brushes are small articles of wealth and I don't know of any crazy socialist that would want to help Mr Bush own and manage his tooth brush or vice versa. What socialists do object to is that the big tools of production and distribution shall be privately owned and that the owners should live and amass wealth out of all human reason . . . and live at ease and squander the surplus in riotous living.[86]

Hopkinson stated that the teachings of the Socialist party were based firmly on the theories of Marx and Engels, not on the utopian ideas many socialists of an earlier day had held. In response to charges that there was no single clear socialist philosophy, he declared, "The man who says there are as many kinds of socialism as there are socialists simply shows his ignorance of the greatest movement since the dawn of history."[87] Hopkinson held to the Marxist notion of two classes in an industrialized society engaged in constant struggle. He wrote: "There is no identity of interests between the two classes and cannot be. This constitutes the class struggle and will exist as long as one class produces wealth and another owns it. When the worker receives the full product of his toil, through the social ownership of the means of production and distribution, the class struggle will cease."[88] In Hopkinson's view, farmers were clearly on the side of the toiling workers cheated of the fruits of their labor, although many urban Socialists persisted in seeing farmers as petty capitalists doomed to extinction. The issue of ownership of farmland was one the Socialist party struggled with and was a frequent topic of questions sent in to the *Appeal to Reason*. The party's farm program, finally adopted after much debate in 1912, advocated a number of programs popular among family farmers, including state-owned grain elevators and warehouses; state-issued insurance and low-interest loans; government encouragement of farmer cooperatives for buying and selling; an exemption from taxation of farm improvements and implements up to $1,000; and steep taxes on land held for speculation or by absentee landlords. The party emphasized that the public interest in land use took precedence over private titles, but that those occupying and using land for farming should have the right of continued occupancy for themselves and their children—although not outright title—once they had paid for the value of the land. The practical implications of Socialist land policy as explained to questioning farmers by the editors and guest writers of the *Appeal to Reason*, was that small-farm owner-operators would continue to hold their land, while tenants would gain long-term rights to the land they farmed, and land held by speculators would be returned to the public domain and opened to landless laborers for either individual or collective farms.[89]

However much the Socialist party might continue and build on Populist positions, the fact still remained that Lewis county Socialists did not have the social organization among the English-speaking population to connect its political program directly to the rural community and mobilize the electorate. The flourishing Granges dominated rural public life. They included some Socialists and embraced political reform, but remained strictly nonpartisan. In its separation from the mainstream of community life, the Socialist party was like the other political parties of the period, tied to a national and state leadership and agenda, but without the strong cultural connection and forum for debate and consciousness-building that the Populists had had in the Farmers' Alliance.

The one place where Socialist organizers might have had a direct connection to rural social life was among the Finnish community centered to the south of Winlock. The Finns had begun arriving in the area in the spring of 1903, when three Finnish men came from the coal mines of Carbonado, Washington, looking for farmland. They bought logged-off land from the J. A. Veness Lumber Company, one of the area's largest employers, and returned that May with their families. Word spread quickly through Finnish communities in mining towns throughout the West. By the time of the 1910 census, seventy-five Finnish families and over forty single male boarders lived in the vicinity of Winlock. Most of the families lived on farms, but only forty percent of the household heads were farmers. The rest worked in the sawmills and logging camps, saving up the cash to improve their stump-covered land.[90]

Rather than joining existing community associations in the early years, the Finns formed their own groups. The Finnish Community Association built a hall southwest of Winlock in 1907 which became the center for a number of activities, including a band, an orchestra, choirs, athletic teams, plays, and a military drill team which conducted its operations in Russian since most of the men had served in the Russian army. A Kaleva, or fraternal lodge with insurance and social functions for its members, was organized in 1908. Finnish Lutheran, Evangelical Lutheran, and Congregational churches were also established. Winlock-area Finns built strong bonds of community and support nourished by frequent social activities, just as did their more Americanized Grange neighbors.[91]

Socialist organizers tapped into this strong community network. Socialist speakers and performers, presumably Finnish themselves, also used the Finn hall, according to the memoirs of Jenny Pelto. While in some midwestern communities "church Finns" split from "Red Finns" and supported dual halls, no such split is evident here. Most national studies of immigrants and politics have found Finns to have been the most consistently radical of any immigrant group during this period, and the Finns had the largest foreign-language affiliate with the Socialist party despite their relatively small population. Factors social scientists cite to explain this radicalism include the degree of rural proletarianization in the homeland prior to migration; the hazardous industrial jobs many took upon arriving in the United States, especially in mining; a strong Finnish socialist leadership and network of speakers; and a close-knit community, which the Winlock Finns typified. Washington ranked fifth of all the states in Finnish population early in the twentieth century, and a Finnish community also flourished across the Columbia River in Astoria. The Finnish population was dense enough in the area to support a foreign-language socialist organization, but little evidence has been left in the historical record. Local newspapers never mentioned socialist organization among the Finns, even during World War I when both immigrants and radicals were viewed with great suspicion. Rather, Finns were portrayed in the press as model citizens. We do know, however, that the three precincts where most of the local Finns lived, Ainslie, Veness, and Prescott, had among the highest Socialist party votes.[92]

The 1912 election was a complicated affair. Only the Republican party ran a stand-pat conservative ticket. Roosevelt's Bull Moose party and the Democrats with Wilson offered two versions of Progressive reformers, while the Socialists offered their leader, Eugene Debs, in his strongest showing. The Prohibition party also ran a national slate. All but the Prohibition party fielded local slates as well in Lewis county. As in the 1890s, voters organized campaign clubs, particularly for Wilson and Taft, and the perennially weak Democrats sought cooperation with other reform tickets—in this case with the local Bull Moose party. The *Bee Nugget* reported that the Democrats and Progressives quietly agreed

Table 9. Percentage of 1912 vote for presidential candidates

Precinct/town	Republican	Democrat	Socialist	Progressive	Prohibition
Eastern farming precincts					
Alpha	12.7	22.8	25.3	38.0	1.3
Cinebar	30.9	20.0	32.7	10.9	5.5
Ethel	22.8	16.5	30.4	27.9	2.5
Ferry	38.2	32.4	11.7	17.7	0
Granite	12.9	22.6	48.4	12.9	3.2
Salkum	12.9	24.8	29.7	29.7	3.0
Western farming precincts					
Cowlitz	34.8	33.9	8.0	19.6	3.6
Drews Prairie	30.4	34.8	21.7	13.0	0
Eden	30.2	5.8	30.2	33.7	0
Economically mixed precincts					
Ainslie	17.6	10.8	36.5	27.0	8.1
Napavine	23.8	32.3	8.9	23.8	11.2
Prescott	34.3	14.3	30.0	21.4	0
Salmon Creek	29.4	9.8	12.0	46.7	2.2
Stillwater	45.7	21.7	4.4	28.3	0
Veness	20.2	14.3	39.3	25.0	1.2
Incorporated towns					
Little Falls	48.8	25.4	1.9	23.0	1.0
Toledo	35.8	25.3	11.7	24.7	2.5
Winlock	30.2	33.4	12.6	21.1	2.6

before their county conventions not to run strong candidates for the same offices so as not to split the opposition. Still, it was a fairly quiet campaign season locally. Almost everybody supported low taxes and some reform. Consequently, virtually no candidate won an outright majority, and voters freely split their tickets.[93]

No single clear pattern emerges to explain the election results in the eighteen precincts in the study area (see Table 9 and Map 5). The predominance of agriculture or logging, rate of group affiliation, percentage and distribution of immigrants other than Finnish, value of property holdings, and rates of mortgages all varied considerably within the various groupings. Some interesting patterns are apparent, however. The precincts that voted most heavily Socialist were of two distinct types. One group, including Granite, Ethel, Cinebar, Salkum, and Eden Prairie, was heavily agricultural, had high rates of persistence between the 1900 and 1910 censuses, and middling-to-high property values that had improved dramatically in the previous decade. They also had relatively high rates of Grange membership. The people here were doing pretty well financially

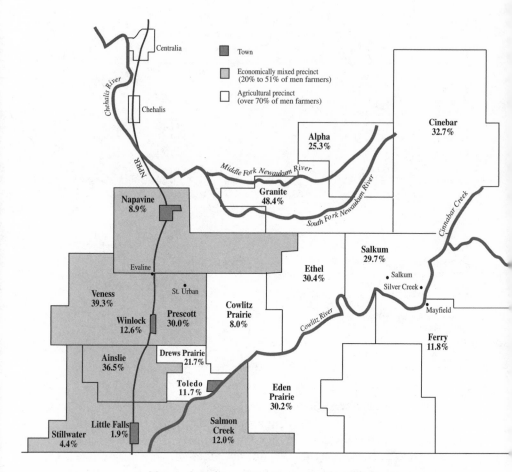

Map 5. Socialist vote in 1912 presidential election

and had a high degree of commitment to their communities, but were hampered by their relative remoteness from markets and transportation routes. None had direct access to a town or rail service. The Socialist doctrine of public ownership and management of the means of distribution would have had a strong appeal to these farmers.

The second pro-Socialist group included Ainslie and Veness, and to a lesser extent Prescott. These were the precincts with the bulk of the Finnish population. All three were distinctly mixed economically, with large numbers of men employed in timber and other industries as well as agriculture. Most families lived on farms, but many of them were newcomers and their land was unimproved. Certainly it is not surprising to find the Finnish precincts in the Socialist camp. Most Socialist voters in the study area were native born, however, and the Finns in fact resembled other Socialist voters in important ways.[94] All of the Socialist precincts were characterized by strong social bonds and a certain economic marginality tempered by the potential for better times in the future. The people of these precincts were still building up their farms. They had not yet reached the relatively comfortable prosperity of the farmers on Cowlitz Prairie or around Napavine, and they identified themselves as producers rather than as business people.[95]

The four towns in the study area with predominantly working-class populations did not give strong support to either the Socialists or the Progressives, although Winlock returned between ten and seventeen percent of the vote to Socialist candidates from 1908 to 1916. Both the Republicans and the Democrats ran well in Winlock, Napavine, Little Falls, and Toledo. The election of 1912 was a disappointment to the Republicans, who ran far behind their 1908 totals, and an exceptionally good one for the Democrats who won pluralities in Winlock and Napavine and gained ground in Little Falls and Toledo. Ferry, which was largely agricultural but included enough of a town center to have over twelve percent of its men classified in business or professional occupations, moved toward the voting patterns of the other towns in the twentieth century, rather than continuing with the third-party leanings of the rural precincts surrounding it.

The election returns from 1912 defy simple categorization or explanation. People who lived near a town center, whatever their own occupation, tended to vote for one of the two major parties rather than risk moving outside the established system. Finns largely supported the Socialist party, but precincts with concentrations of Germans, Norwegians, and Swedes were scattered across the political spectrum.[96] Farmers and aspiring farmers were more likely to vote Socialist than people from working-class or business families. Although there were large numbers of timber and other workers in Lewis county, including in some of the precincts with the highest Socialist vote, there is little evidence to suggest strong labor influence. Granges never discussed conditions in the local mills and logging camps, even though many members had experience in the industry and they expressed support for more distant and impersonal workers. The Finns, despite the background many shared in mining communities and their high degree of organization generally, apparently did not form labor unions among their timber workers. The IWW as yet had made little headway in its efforts to organize the timber industry, and the craft unionism that prospered in Chehalis and Centralia was of a fairly conservative variety. The alliance between farmers and industrial workers existed primarily at the state level—it had not arisen from grass-roots concerns. The timber workers who voted Socialist in Veness and Ainslie were aspiring farmers, not committed to a class struggle with mill owners. Lewis county Socialists sought significant reform of the political and economic system and championed a revolution in values. They wanted a society where labor and production for human use were valued more highly than the making of money. These were the same goals local farmers championed when they first voted Populist in 1892. Twenty years later much had changed, but the basic structures of society and the everyday problems had not.

Politics had changed, however. Not all former Populists voted for the Socialist party, even if they remained committed to their ideals. The other parties, too, had now accepted the necessity of reform. The Democrats after unsuccessfully nominating Bryan three times, still continued to run on a platform committed to ending the political and economic dominance of big business. In Washington state and elsewhere, many Republicans also advocated Progressive reforms, even if the national platform remained conservative. In the relatively quiet and amicable political competition of the early 1910s, choices were not necessarily clear-cut. Whether Socialists in this context could rightly be labeled as radicals is questionable. It must be emphasized, nonetheless, that substantial numbers of rural voters chose the Socialist party over the reform tickets offered by the Democrats and Progressives. Surely that decision was not made without deliberation.

Lewis county Granges both responded to the appeals of their state leadership and formulated their own issues. In this respect they were similar to the Farmers' Alliance locals, but there was a major difference. In the 1890s, the economic and political critique and the overall reform agenda were developed largely at the national level; contributions came from organizations in many regions of the country, and Washington state was a relative newcomer. In the early twentieth-century, the national reform effort was more scattered, and leaders of the State Grange and Federation of Labor were often far in advance of their largely conservative national counterparts. Consequently, state leadership was more influential at the local level in the twentieth-century, and state issues had a greater prominence than national. While twentieth century agrarian protest in Lewis county was thus not so closely tied to the success or failure of a national movement, it also lacked some of the vitality and broad focus that diversity and a national coalition had brought to the Alliance and the Populists.

After the disappearance of the Farmers' Alliance and the disintegration of the People's party, rural Populists were forced to find new ways to achieve their goals. In the South, sharp divisions of race and land ownership and the reestablishment of white planter dominance effectively buried the Populist spirit.[97] However, Lewis county's more egalitarian economic structure and the successful organization of the Grange in many rural neighborhoods allowed the practice of democratic participation and the pursuit of economic justice to continue to flourish. The partisan expression of those beliefs was not as clear in the early twentieth century as it had been in the 1890s, but the commitment to a democratic process remained strong. Rural people met regularly as brothers and sisters, took responsibility for improving their local communities, educated themselves about and debated state and national concerns, lobbied for political reform, and recognized their kinship with others as diverse as Colorado miners and Mexican campesinos. The Grange also continued to give farm women new opportunities. While accepting many common notions of sex differences and proper roles, the Grange also acknowledged women's contributions to the family farm economy and fought to establish their full rights of citizenship—to vote, to discuss political and economic issues, and to provide leadership in their communities.

Political reform was only one aspect of Progressivism to reach into and be shaped by Lewis county's rural neighborhoods. Farmers did not wait for the results of political reforms or government action to strengthen their economic positions. They took a number of actions individually and collectively to improve their farming techniques and gain control of marketing systems. At the same time, new government initiatives in the form of scientific agriculture and the Extension Service also came to the countryside in the early twentieth century. The strong political consciousness nurtured in the Grange helped shape farmers' responses to these new opportunities as well.

ℭ𝔩 6 𝔩℈

Specialization and Cooperation: Agricultural Change in the Early Twentieth Century

While Lewis county farmers continued to nurture community ties and build on Populist political traditions in the Grange, they also found new ways to achieve economic independence. Most farmers raised a variety of crops for home consumption and the market in both 1900 and 1925, but specialization increased dramatically during this period, making dairy products, poultry, and, to a lesser extent, fruits and berries the dominant market crops in Lewis county and much of western Washington by the early 1920s. During the same period, farmers formed increasingly larger-scale cooperatives to process and market their produce. These cooperatives dominated the agricultural economy of the region in the 1920s and well beyond, giving farmers in large measure the financial security they had long sought. Although membership overlapped significantly, the cooperatives were organizationally separate from the Grange and usually had different leadership as well, allowing them to seek support from and work amicably with businessmen's clubs in towns despite growing political tensions.

Government involvement in the local economy also expanded in the 1910s, as the Extension Service promoted scientific agriculture and a particular view of rural life. In many ways, the Lewis county farmers of this period represented the ideal portrayed by the Extension Service. They studied market conditions and the findings of agricultural scientists. They used top-quality, purebred stock, followed standardized grading to assure consumer confidence, bought and sold through cooperatives in order to maximize their profit, and worked closely with town businessmen. At the same time, Lewis county farmers accomplished the changes in agriculture through their own efforts and on their own terms. Their general response to the Extension Service ranged from cautious to hostile, and they spurned the government-endorsed and politically conservative Farm Bureau, instead remaining loyal to the Grange and the producers' alliance. Just as in the 1890s, the desire for a decent return on honest labor and control of their own affairs lay at the heart of the farmers' organized efforts. While they were quite willing to take advantage of what they saw as useful offerings of the Extension Service and business community, they insisted on keeping control in their own hands rather than ceding it to outside "experts." Farm men and women also differed from the Extension Service ideal in continuing to hold to an ethic that viewed the farm wife as a working partner with her husband, not a model of domesticity.

These changes occurred as frontier settlement patterns gave way to characteristics of more stable regions.[1] The population of the heavily agricultural eastern precincts in the study area had increased by 25 percent during the 1890s, and by another 31 percent by 1910. Strong growth continued in Ethel and Salkum precincts the following decade, but Alpha and Ferry actually declined in population and Cinebar changed little. Despite population growth between 1900 and 1910, the character of these precincts stayed much the same. (See Figure 8.) They remained overwhelmingly agricultural, with 80 percent of men farmers, and 87 percent of women described as housekeepers. About half of the adults counted in the 1900 census could be found in the 1910 census as well, and a growing percentage had lived all or most of their lives in Washington. Newcomers continued to be largely native-born, with some immigrants from Scandinavian and German-speaking countries. In 1914, a major lumber mill opened in Granite, creating the boom town of Onalaska in that previously rural precinct. The surrounding countryside retained its agricultural orientation.[2]

This period is usually considered a golden one for American farmers, with prices for farm produce high and indebtedness low.[3] Indications are that Lewis county shared in at least some of that prosperity. Farms of these eastern precincts showed profitability during the first decade of the century, although not enough for a family's way of living to change much. Among residents whose tax and census records could be linked, real property assessments increased by over 45 percent between 1900 and 1910. Length of residence was a major factor in this increase. The large core of stable property

Figure 8. Threshing machine and engine on Griel farm, Ethel, Washington. Photograph courtesy of Lewis County Historical Museum, Chehalis, Wash.

owners had had the time and commitment to invest substantially in their farms (see Table 10). Newcomers on average owned property worth less than long-time residents and were also more likely to have mortgages or to rent their land (see Table 11).[4] On the other hand, assessed values for personal property changed little, and the percentage of adults holding taxable personal property actually declined between 1900 and 1910. Farmers evidently plowed their profits directly back into their farms rather than investing in consumer goods. A U.S. Department of Agriculture study of western Washington farms conducted in the early 1920s found that farmers who had settled before 1898 put all their earnings back into their farms until the late 1910s. Only in the post-World War I years, farmers reported, did they begin to spend money on such items as travel, education for their children, or stocks and bonds.[5]

While the eastern precincts remained overwhelmingly agricultural, in the rail corridor town population growth far outpaced rural growth after the turn of the century. The percentage of households headed by a farmer decreased from 60 percent to 36.6 percent between 1900 and 1910 in the precincts surrounding Winlock, al-

Table 10. Average value of real property and rate of ownership by length of residence for all precincts, 1910

Length of residence	Number of people assessed	% of population assessed	Average assessment in dollars
Arrived before 1900	414	39.3	960
Moved within area	81	32.5	793
Arrived after 1900	575	18.2	565
All real property owners	1070	24.0	735

Table 11. Farm ownership of households living on farms by length of residence for all precincts, 1910

Length of residence	Number of households	Farm		
		Owned	Mortgaged	Rented
Arrived before 1900	280	86.1%	8.9%	5.0%
Moved within area	44	70.5	20.5	9.1
Arrived after 1900	451	64.1	24.4	11.5
All households on farms	775	72.4	18.6	9.0

though the actual number of farming households did increase. Three of the precincts—Cowlitz Prairie, Drews Prairie, and Eden Prairie—remained primarily agricultural in the early twentieth century, with over three-quarters of households headed by a farmer and over 70 percent of all men listed as farmers in the census. In six other Winlock area precincts at least one-third of the households were headed by a farmer: Stillwater, Salmon Creek, Prescott, Veness, Ainslie, and Napavine. (The town of Napavine was not separated from the surrounding rural district until the creation of Emery precinct in 1914.) Two of the precincts, Veness and Ainslie, immediately west and south of Winlock, more than doubled in population between 1900 and 1910 as Finns and other newcomers purchased the logged-over land for farms.

Many of the Finnish immigrants as well as sons of long-time residents continued the practice of earning cash to buy and improve farmland by working in the mills and logging camps. After J. H. and Lizzie England bought a hundred-acre farm near Evaline in 1904, J. H. walked to his mill job in Napavine and Lizzie kept the farm and cared for their infant son. They joined Hope Grange in 1905.

John Schafer was only able to come home on weekends from his mill job, while his wife, Emma, and their four daughters raised corn and cucumbers and kept a small herd of cattle. Indeed, the 1910 census indicates that many more families lived on farms than had farmers as their primary breadwinner. In fact, in all but the four town precincts, the majority of households lived on farms even in those precincts where the majority of men held off-farm jobs.[6]

Despite the expansion of the farm population, the number of acres of total farmland changed little in Lewis county between 1900 and 1920. By 1920, the size of the average farm was seventy-five acres, less than half what it had been in 1890.[7] Movement to new lands was prevented in part because so much acreage was tied up in federal reserves on the flanks of Mount Rainier and Mount St. Helens, and increasingly in the twentieth century by the holdings of timber companies, including Weyerhaeuser. The problem was also with the land itself. The first white settlers believed that soil that could grow a two-hundred-foot-tall Douglas fir could grow anything, but in fact the soil was usually too acidic to make good farmland.[8] An even bigger hurdle was in clearing the land. Even after the trees had been harvested, a prospective farmer faced the enormous task of removing the giant stumps and tangle of branches, tree tops, small trees, and undergrowth the loggers had left behind. A U.S. Department of Agriculture bulletin published in 1924 estimated that clearing one acre of logged-over land for farming in western Washington required fifty days of human labor, thirty-four days of horse labor, and 205 pounds of explosives. The authors determined, not surprisingly, that the cost of clearing far exceeded the value of the land. Rather than trying to cultivate the land immediately, a Farmer's Bulletin published in 1911 recommended burning the slash in late summer when it was most likely to be dry, sowing a carefully chosen mixture of grass seeds in the resulting ashes, and once a pasture was well established, turning it over to cattle and goats—which could be counted on to maintain the pasture and eat new weeds and brush, unlike horses or sheep. Once the stumps were well rotted, they could be dug out and the ground plowed for cultivation.[9] The difficulties of opening new land, the subdivision of the large land grants of the early white settlers for the next generation, and the move to poultry and dairy farming all contributed to the rapid decline in average farm size in the early twentieth century.

Despite these difficulties, Lewis county farmers enjoyed the ad-

vantages of fertile land in the river bottoms and prairies, abundant
rainfall, moderate temperatures, and full land ownership. Therefore, unlike their counterparts in many areas of the country who
suffered the consequences of overworked land, drought, monocul-
ture, and structured indebtedness, they were able to grow a mixture
of crops to feed themselves and their stock, and to respond to
changing market conditions with some flexibility. Those who bought
logged-over land began raising poultry and fruit, which would turn
a profit even on farms with few acres cleared.[10] While professionals
in the Country Life Movement condemned American farmers as un-
sophisticated, ignorant followers of tradition, Lewis county farmers
continuously displayed a self-conscious determination to remain independent and to prosper by adapting their crops and techniques
to the demands of the international market, improving technology,
and their natural environment.[11]

Cooperative marketing of specialized crops was perhaps the strategy
most consistently adopted by Lewis county farmers to strengthen
their economic position. Hops farmers were the first locally to attempt organization. Most of the hay, potatoes, vegetables, and meat
that farmers raised during the 1890s was consumed locally and
could be sold at relatively stable prices, but the price for hops was
determined by an international market that varied wildly. Farmers
responded to the hop market by adjusting the number of acres
grown, but a strong market in the spring by no means guaranteed
prices would remain high by the time the crop came in. Thus hop
raisers turned to organization to gain some control over marketing
their crop, but early efforts proved too limited in scope to be successful. In November 1891 two hop growers, John Dobson and
Charles Wuestney, a founding member of the Farmers' Alliance in
Lewis county and soon to be editor of the *Advocate*, called a meeting
in Chehalis to organize a county hop-growers association. The meeting was well attended and appointed a committee to draft bylaws,[12]
but the organization seems to have died soon after. In 1894, a second association was formed with different leadership. They set up a
sample room in Chehalis, hoping to attract international buyers to
view the local crop. The sample room met with only limited success,
but the organization did persist for a second year. In 1895, a second
year of disastrously low hop prices, the growers' association agreed
to lower the pickers' pay from $1.00 to $0.75 per box, causing quite

an uproar. The *Advocate* was not very sympathetic with the growers and certainly found nothing Utopian about the current situation, stating, "The contention of the grower, that hops cannot be sold at the present price and pay $1.00 is true, but on the other hand, can the laborer afford to pick for seventy five cents."[13] From then on, hop growers apparently gave up the attempts at organization. They had been unable to overcome the low prices of the international market with such a small-scale organization, and when prices rallied, were able to do well individually. Ultimately, the outbreak of war in Europe, followed by the passage of Prohibition, eliminated the bulk of hops cultivation in Washington by 1920.[14]

The berry and tree-fruit industry in many ways paralleled the hops business—in fact, many discouraged hop raisers turned to growing fruit. A number of similarities facilitated the shift: both crops were raised primarily for the market, although some home consumption of fruit was also possible; like hops, fruit could be raised profitably on almost any size plot when prices were good; local women and children and migrant labor picked the crop; and fruit and berry raisers frequently tried organization as a way to improve their market position. In contrast to hops, though, fruit was still a growth industry in the 1920s. Because consumption of fruit was widespread and canning could be done locally, farmers' efforts at cooperation had a greater chance of success.

Fruit growers in the county made their first attempt at organizing in 1892. F. A. Degeler of Newaukum and David Motter of Cowlitz Prairie, both active in the Farmers' Alliance and Populist party, were elected president and vice president of the short-lived organization. Motter went on to become horticultural inspector for the county in 1901, writing long letters to the newspapers about such topics as root lice and wooly aphids.[15] In late 1910, a far more ambitious organization was attempted by a group of fruit growers and Chehalis businessmen. They established the Southwest Washington Fruit Growers' Association, hoping to build their own cannery and box factory, prepare spraying compounds, and sell spraying machinery. They were inspired by a similar undertaking in Puyallup and Sumner, site of the Western Washington Agricultural Experiment Station.[16] By 1914 a cooperative cannery was operating in Chehalis, with over five hundred stockholders. It reportedly did $10,000 worth of business its first year, $125,000 in 1917, and undertook a major addition in 1918 that was expected to double its capacity and increase its work force to two hundred women.[17]

The cannery business continued to boom until 1921, inspiring farmers and entrepreneurs in several other parts of the county to organize smaller-scale ventures. Canneries were opened in Randle in 1919 and Pe Ell in 1920.[18] Then the nationwide agricultural depression of 1921 hit fruit raisers hard. In May of that year, the *Bee Nugget* declared there was a large surplus of canned goods in the United States, and the Chehalis cannery urged growers to can their berries at home. By late summer, the situation had improved, the stored 1920 crop began moving, and the cannery once again started accepting new produce.[19]

The push to organize continued. Berry growers, perhaps feeling their particular concerns were overwhelmed by the volume of tree fruit and stung by losing their early summer crops in 1921, formed their own association in 1923. They affiliated with the North Pacific Cooperative Berry Growers, headquartered in Seattle, and leased a plant in Chehalis to handle their produce. Farmers from as far away as Winlock trucked their produce to Chehalis after dark and marketed through the North Pacific Berry Growers. Other farmers around Winlock and Toledo, rather than join the Seattle cooperative, bought into the locally owned Cowlitz Produce Company. In the first three weeks of June 1923, the *Winlock News* reported, Cowlitz Produce shipped eleven rail carloads of strawberries from Winlock and seven from Centralia, with farmers receiving "exceptionally high" prices. They were also selling directly to Winlock merchants.[20]

That same year businessmen and farmers in the Winlock area were excited by the visit of an entrepreneur named Searles, who proposed opening a cannery in that town. Farmers from Toledo, Evaline, Napavine, Cougar Flat, and St. Urban all agreed to put up stock; plans were drawn and a local contractor hired. At that point Searles withdrew, however, claiming too many other projects already underway, and the effort was apparently abandoned.[21]

In many ways, the fruit industry of Lewis county appears to represent the ideal advanced by the Extension Service and the Farm Bureau in the 1920s: farmers and townspeople, businessmen all, working together to advance their mutual interests. Certainly the interests of the two groups did coincide in major ways. Town businesses relied on the trade of prosperous farmers, and both groups benefited financially when farmers had a virtually guaranteed market for their produce. The editorial slant of the county newspapers endorsed this vision, promoted joint ventures by farmers and businessmen, probably inflated their successes, and ignored tensions and failures. Some hints of tension do come out in the press accounts, however. Organization on the part of the farmers was inspired at least in part by opposition to business as it was being conducted. By 1912, the Pacific Fruit and Produce Company apparently had a near monopoly on the fruit market of the region. Like wheat and cotton growers in other parts of the country, Lewis county fruit raisers were not satisfied with the prices they were being offered and believed the profits from their labors were ending up in other hands. Organizing among themselves and supporting the development of local canneries gave them the strong probability of receiving higher prices. Farmers were also unhappy with local merchants, who often bought produce not from local farmers, but from the large distribution companies that could probably assure a more steady supply and more varied goods. But the farmers believed that Lewis county people should buy Lewis county produce, and at a fair price.[22] They were in the business of farming not to make huge profits like the large corporations, but to earn an honest and sufficient living in a way that benefited the whole community. The tensions raised by these issues always underlay dealings between farmers and town merchants.

Dairy and poultry farming became the big agricultural businesses in Lewis county in the early twentieth century, with Winlock proclaiming itself the "Egg and Poultry Capital of the World," and dairy farmers forming a cooperative that still thrives today and markets dairy products internationally. Lewis county farmers began attempts to organize the dairy business in the early 1890s. The *People's Advocate* reported a number of local small-scale efforts. In 1892, the Farmers' Alliance planned a creamery for Eden Prairie that apparently was never built. Silver Creek farmers formed a stock company in 1893 in response to a proposal for a creamery there by an E. H. Garber from Pennsylvania. However, Garber refused to build until he had the money and pledges of milk, and the farmers would not go any farther until they saw a building, and so the deal fell through. Creameries apparently were built in Chehalis, Mossyrock to the east of Silver Creek, and Boistfort to the west of Winlock, in 1896 and 1897.[23]

In 1899 farmers of Cowlitz and Eden Prairies responded eagerly when Charles and Ernest Kaupisch proposed building a creamery in Toledo. Capital stock of $600 was raised by selling shares of one dollar. The Cowlitz River Creamery, as the new venture was called, was not strictly a cooperative, but a corporation whose stock was owned by its entrepreneurial organizers, the farmers it served, and other local investors. Washington state law did not recognize farmers' cooperatives as a separate kind of legal entity until the passage of the Cooperative Marketing Act of 1921. Before that time, cooperatives organized under private corporation law, just as any corporation would. What distinguished a cooperative was that only member farmers bought shares—in a dairy cooperative, usually one share for each cow milked—and then elected a board and hired a manager.[24] The Kaupisch Creamery stockholders included many of the dairy farmers who used its facilities, but also representatives of other prosperous farming families in the area and Toledo businessmen. The president of the board was a former member of the Farmers' Alliance. The creamery opened April 10, offering "the highest prices" for milk and selling milk cans at a low cost to farmers. Two years later, the Kaupisch brothers' operation had expanded to the point that they opened a second station at Adna. That year the *Bee Nugget* reported that beef cattle were so scarce in the county that meat was being supplied from Montana, but the dairy business was booming. A major new creamery was being built in Centralia, and smaller ones elsewhere. "The dairy industry of Lewis county is a paying one if conducted along proper lines. During the past year many farmers have been convinced of this fact," the *Bee Nugget* declared.[25]

Dairying remained a paying business. In 1913 one Toledo-area farmer estimated earning $100 per month from fifteen dairy cows.[26] By that time, Lewis county farmers had a number of choices for marketing their milk. One option was to sell to private companies. The biggest was the Pacific Coast Condensed Milk Company which operated a plant in Chehalis. It is clear, however, that many farmers preferred to control as much of the marketing apparatus as possible themselves, for they continued to work for cooperative associations and local corporations in which they owned stock. The Toledo creamery was producing 40,000 pounds of butter annually by 1913, and another $6,000 worth of Toledo-area cream was being sent each month directly to Seattle. The Winlock Cooperative Creamery was established in 1906 when farmers took over an already existing private creamery. It expanded rapidly under cooperative management, producing over 20,000 pounds of butter in May 1913.[27] The Lewis County Cooperative Creamery built a plant in Napavine in 1912 serving the farmers from Napavine east to Mossyrock, including those of Ethel, Salkum, and Silver Creek. The farmers had spent some months meeting and discussing their options. They considered building at either Napavine or Chehalis, and possibly affiliating with a recently formed cooperative at Boistfort. At one point, they collected subscriptions in response to an offer for a creamery from two Seattle men who were later determined to be crooks. In the end, the dairy farmers took advantage of a bonus offered by the Citizens' Club of Napavine, formed their own cooperative, and built in that rail-served town. By June 1913, after ten months' operation, the Lewis County Cooperative Creamery had produced over 150,000 pounds of butter and paid out $44,000 to its farmers.[28]

Of these relatively small organizations, the Winlock Cooperative Creamery was by far the most successful. By the early 1920s it handled virtually all of the milk produced in the vicinity of Winlock, and its 1924 sales exceeded $167,000. The Kaupisch creamery of Toledo had gone out of business by the 1910s, but in 1918 sixty-seven farmers of that area formed the Cowlitz Valley Cheese Association, began sponsoring an annual Cheese Day in Toledo, and sold $35,000 worth of cheese in 1924. The Lewis County Cooperative Creamery struggled with debt through most of the 1910s, but was running a positive balance in 1919. By the mid-1920s, it had merged with the much larger Lewis-Pacific Dairymen's Association.[29]

The formation of the Lewis-Pacific Dairymen's Association (LPDA) represented a whole new level of cooperative organization for area farmers. Begun in 1918 as the Lewis County Dairy Association, it claimed to represent 50 percent of the cows in the county by January 1919 and launched a membership drive with the goal of reaching 90 percent representation. In May 1919, the organization formally incorporated as the Lewis-Pacific Dairymen's Association, expanding into neighboring Pacific county.[30] The LPDA quickly affiliated with the United Dairy Association of Washington, which had been organized at about the same time. J. A. Scollard, a dairy farmer near Chehalis, was a major organizer and president of both the LPDA and the UDA. The UDA was composed of five county-level dairy cooperatives that covered most of western Washington except for the Seattle-Tacoma and Portland-Vancouver areas. Each member association marketed dairy products within its own county, while the UDA handled wider distribution and assured standardized products. Collectively, they adopted the brand name Darigold.[31]

The LPDA launched an ambitious building drive, constructing a utility plant in Chehalis that could produce butter, cheese, and powdered and condensed milk, and could shift among products as the market varied. In 1921, the $175,000 plant opened in defiance of collapsing agricultural markets. It received 20,000 gallons of milk its opening day, expected 80,000 gallons per day by the end of its first week, and had the capacity to handle 200,000 gallons a day. In 1923 the LPDA plant took in nearly 22 million pounds of milk for which it paid $709,000 to its 853 members. Both the LPDA and the UDA continued to prosper, and today the Darigold Corporation, as it is called, remains strong.[32]

The 1919 bylaws of the LPDA required that members be residents of Lewis, Pacific, or adjoining counties, be engaged in dairy farming, and purchase a $10.00 share for each cow owned. (The Winlock Cooperative, in contrast, charged members $5.00 per cow.) Each member had one vote regardless of the amount of stock held. Trustees were given full power to run the corporation during their three-year terms, for which they received no compensation. Trustees were also empowered to impose rules on their members regarding feeding and care of livestock, the storing and transporting of milk, and standards of cleanliness and quality.[33]

The success of the LPDA affected both private companies and the other dairy cooperatives. Carnation was the only major private dairy company operating in Lewis county when the LPDA began. In October 1920, after construction of the LPDA plant had already started, the *Bee Nugget* reported the collapse of the dairy market in New York. The following issue announced that Carnation was cutting its prices to farmers from $2.90 to $2.50 per hundred pounds of milk. In early November, Carnation laid off 65 of its 75 Chehalis workers and stopped condensing milk.[34] Work on the LPDA plant continued, however, and new pledges of funds from dairy farmers poured in. Just before the new plant opened, Carnation raised its price to farmers, and during the summer of 1921, the LPDA's first season of operation, Carnation offered $0.30 more per hundred pounds of milk than the cooperative.[35] Judd Bush, an LPDA trustee, accused Carnation of trying to break the new cooperative. Milk prices had always varied considerably, but in this case, Carnation was not only paying local farmers above the LPDA rate, but also reportedly $0.30 per hundred more than the corporation was paying farmers in Wisconsin and Colorado. Farmers had to stick together, Bush and Scollard pleaded, or they would be back at the mercy of national markets and national companies.[36]

But not all the farmers stuck. In December 1921, the LPDA instituted a suit against one of its Pacific county members for selling milk to Carnation in breach of the contract all members signed to sell only to the cooperative. The LPDA succeeded in winning an injunction against him.[37] Then word leaked out that Scollard himself had been selling to Carnation, thus receiving higher prices for his milk. Bush defended his friend, and the LPDA which instituted no action against him, on the grounds that city dwellers and logging camp workers needed the milk Carnation supplied them, and because Scollard continued to pay into the sinking fund for the milk he sold to Carnation—perhaps the more legally relevant fact. Rumors circulated that as many as 200, or one fourth of the LPDA membership, would quit in January when their contracts allowed them to withdraw. The association survived the controversy in good shape, however. Only 34 members actually left, and those numbers were quickly made up with new additions. Scollard remained as president.[38]

The LPDA's relations with the other dairy cooperatives in Lewis county were somewhat more complex than the straightforward competition, and even animosity, toward Carnation. All the cooperatives shared the goal of giving the farmer the greatest possible return on produce and some involvement in the marketing process. At the same time, the more members an association had, the surer its success, the higher its potential profitability, and the greater its power. The LPDA's leaders wanted to handle all the milk produced for market in Lewis county, and through the UDA to unite all dairy farmers in western Washington into a single powerful organization. All dairy farmers stood potentially to benefit. Thirty years earlier, the Farmers' Alliance had advocated just such supercooperatives in order to combat the strength of the large corporations, and to give farmers a fair return on their labor and consumers fair prices. The push for ever larger and more successful organizations, however, came at the price of local control. Many dairy farmers in Lewis county were already in smaller-scale cooperatives where their neighbors or they themselves served as trustees, and where the paid manager was a trusted friend. These smaller organizations had more flexible rules, allowing members to sell elsewhere and nonmembers to sell to the cooperative. The Winlock, Lewis County, and Cowlitz Valley associations relied on loyalty and the quality of service offered, rather than on binding contracts, to hold members.[39] Dairy farmers had struggled to build their cooperatives and to hold onto them during adverse times. For many, local control was as important an issue as level of profit. They were understandably reluctant simply to be absorbed by the LPDA.

The Lewis County Cooperative Creamery was the most vulnerable of the three, with a shaky financial history. In March 1921, T. J. Long, the creamery manager, announced they would continue operation after the new LPDA plant opened, despite rumors of merger. Within two years, however, the LCCC had been incorporated into the LPDA.[40] The Winlock Cooperative Creamery, on the other hand, continued strong and profitable, building an extension in 1923. Both it and the Cowlitz Valley Cheese Association finally sold out to Darigold, as the LPDA-UDA was widely called, in the mid-1940s when the larger organization began picking up whole milk from the farm, saving farmers the work of separating and hauling.[41]

The LPDA might have been on a scale visualized by the Farmers' Alliance a generation earlier, but its leaders advocated a very different philosophy of political economy. LPDA organizers emphasized their role as businessmen rather than as members of the "producing classes," and it remained clearly separate from the Grange, which dominated rural social and political life. The founding meeting of the Lewis County Dairy Association, precursor of the LPDA, was held in the hall of the Chehalis businessmen's association, the Citizens' Club. Scollard and Bush wanted no connection to the radical roots of the farmers' cooperative movement, a position heartily endorsed by the *Bee Nugget.* Scollard, in a later address to the Citizens' Club, proclaimed farmers the natural allies of small-town businessmen, not of laborers.[42] Judd Bush described the LPDA in 1921 as a totally new phenomenon, the outgrowth of three, not nearly thirty, years of education and organization. In an article for the *Bee Nugget,* Bush

distanced his organization from the Grange and the other existing
dairy cooperatives in the county by ignoring farmers' efforts of previous decades in his account of the rise of the LPDA:

> It is nearly three years since the dairymen of western Washington began
> in earnest the movement to break away from the old order of things and
> go into the manufacturing business for themselves. It has been three years
> of organization and education. Many of the dairymen believed that it
> would be impossible to do anything to relieve the situation and the public
> were decidedly skeptical, some even hostile toward a cooperative move-
> ment among the farmers. But the farmers knew by bitter experience that
> it was either a matter of doing something for themselves or going out of
> the milk business. At first it was believed that by getting together in an
> association and controlling the milk supply the problem could be solved
> but it was soon apparent that some provision must be made for disposing
> of the milk: It was out of the question to throw it away, the producers
> would be the losers and the factories could wait with comparatively little
> loss until the "strike"was over. So a wiser plan was adopted and it was
> decided to build a utility milk plant. This plant could be no little creamery
> or cheese factory, it must be of great capacity, able to handle a big volume
> of milk and turn out a product that would win its way on its merits. Grad-
> ually the visions of the farmers grew. Three years ago to have told Lewis
> County dairymen that they would have to bond themselves to the extent
> of 100s and 1,000s of dollars each in order to "put over" a successful
> factory would have been enough to kill the proposition. Today nearly 700
> farmers of Lewis and Pacific counties are together in a cooperative asso-
> ciation in which they are stockholders to the extent of $10 for every cow
> they own and half of them are holders of the association's bonds.[43]

The denial of a history of farmers' organization and cooperative efforts was particularly important to Scollard and Bush because the LPDA
was formed at a time of increasing political divisions between town and
countryside. The Grange was a powerful organization in the county and
state and would seem a natural ally, but the Grange also continued to
pursue reform politics during World War I, while the conservatism of
the towns hardened. (The political situation will be discussed more fully
in the following chapter.) Too many LPDA members belonged to the
Grange for LPDA leaders to attack it directly, but they carefully avoided
any public association, and made a point of proclaiming their loyalty
to the government and freedom from any political affiliation.[44]

The LPDA and the other dairy cooperatives of the 1910s and 1920s could separate themselves from the other activities of their farmer members because they were strictly economic organizations, without strong community-building components. The overwhelming majority of cooperative members were men, the few women most likely widows. Farm husbands were the ones who joined, as the acknowledged public representative and financial agent for their families. The cooperatives both reflected and reinforced the prevailing sexual division of labor on family farms. This probably unconscious gender conservatism, in turn, allowed LPDA leaders to adopt a conservative political posture at odds with the prevailing views of the membership. Dinners were held annually in addition to stockholders meetings, and farmers lining up at the plants with their carts of milk had some opportunities for socializing that no doubt reinforced their sense of belonging and their loyalty to the association. The cooperatives lacked, however, the regular twice-monthly family meetings of food, fellowship, and discussion that made the Farmers' Alliance, and in the twentieth century the Grange, such strong forces of democratic activism.[45]

In addition to the cooperative movement, a second important element in the economic success of dairying in Lewis county was the upgrading of stock and increased attention to the care and feeding of animals. By the late 1910s, some Lewis county dairy farmers were avid advocates of purebred stock, regular testing of milk for butterfat content, and building silos—all components of successful dairying that the Extension Service aggressively advocated. Most Lewis county farmers, however, were somewhat cautious in adopting new practices and only invested time and money in those that their own experience validated.

The State Agricultural College at Pullman offered the first Farmers' Institute in Lewis county at Chehalis in 1895. Turnout was disappointing, and another was not held until 1900. This second course drew a much larger crowd, and thereafter Institutes were held in various locations around the county with some regularity, covering such practical topics as care and feeding of animals, disease and insect control in orchards, and so forth. Attendance at the Institutes varied considerably, indicating that farm families were quite willing to attend when it was convenient to do so and other activities did not interfere.[46]

Cow testing was one activity strongly recommended by the Extension Service that Lewis county farmers by and large ignored. At the initiative of state officials, a cow-testing association was formed in 1911 for the area around Chehalis to test the butter-fat content of milk. Farmers were advised to weigh the output of each cow as well, and to sell unproductive cows to the butcher. This first cow-testing association lasted about two years. After 1917 when Lewis county commissioners agreed to hire an agricultural extension agent, the agents spent considerable time and effort trying to establish new testing associations. They were frustrated both by the farmers' unwillingness to pay a testing fee and by the difficulty of hiring a tester in a county where lumber camps and sawmills could pay much higher wages.[47] Farmers apparently resisted even very low-cost testing schemes. In 1918, Louise Schaefer of the St. Urban Grange offered to test Grange members' milk at no charge for the following year if they would provide her with the equipment. She renewed her offer a month later, but on both occasions her suggestion was tabled.[48] In 1922, the county agent reported the successful establishment of a new Cow Testing Association, but it only included 35 farmers and 575 cows out of the over 1,300 dairy farmers and 10,000 cows in the county. Still, it was an improvement over the 1918 attempt that only attracted 4 farmers. By 1924 the CTA represented 1,400 cows, and the county agent was determined to continue promoting it.[49]

Purebred stock associations attracted somewhat greater followings. In the late 1910s and early 1920s, a Lewis County Pure Breeders Association, Holstein Breeders Association, Jersey Breeders Association, and Guernsey Club were formed. Most included dairy farmers from around the county, including Alpha, Winlock, and Chehalis. They concentrated as much on socializing as on education, holding picnics and visiting each other's farms, as well as sponsoring lectures and fair prizes.[50] The St. Urban Grange, one year after rejecting Louise Schaefer's proposal for testing cows, set up a stockholders' association and purchased a Holstein bull. They agreed to charge $2.00 to members and $5.00 to nonmembers for the use of his siring services.[51]

Lewis county farmers were not unique in turning to dairying and organization. Farmers throughout western Washington followed a similar course. An Extension Service study of eighty-one dairy herds in four western Washington counties (including Pacific but not Lewis) in 1919 and 1920 described the range of conditions on dairy farms. The typical farm had between ten and twenty dairy cows and about 120 total acres, with about 55 acres devoted to pasturage. On the smaller farms, most of the work was done by family members, who spent the bulk of their time caring for crops—typically three-fourths of the hay consumed by the cows was grown on the home farm, as well as such family food as potatoes and chickens. Only the farms with the largest herds had a full-time worker devoted to the care of the stock. Including the value of family labor and home-raised hay in their calculations, the study authors found a range of $1.81 to $4.56 in the cost of producing one hundred pounds of milk. They determined that forty-nine of the herds made a profit, while thirty-two ran at a loss. Most of this loss was in opportunity cost rather than cash output, however, so many of the families losing money on paper were still getting by. The family might have earned more by working for wages and loaning out their equity in their farms, but, the authors concluded, "it is not uncommon to find families that will live with such frugality under such circumstances that they are actually able to lay up a little money each year. This does not indicate that the cost of producing milk has been figured too high, but it does show the dogged determination to succeed that is possessed by some families."[52]

The poultry industry followed a similar pattern of specialization, cooperative marketing, and rapid expansion, while continuing to remain rooted in the traditions of family farming. Like dairy cattle, chickens were a standard feature of most farms in the 1890s, raised primarily by farm women for both family consumption and the market. But as with other products, once the balance shifted away from home use to production for the market, men took over most aspects of the business. Taina Nelson remembers her mother, Impi Erving, starting the poultry business on their family farm in Veness with money borrowed from neighbors in the early 1910s, but it was her father, Jacob, who was known for his successful chicken ranch in the 1920s, and local histories credit her father, not her mother, with buying the family's first batch of chicks.[53]

Lewis county newspapers began reporting on chicken raising around 1900 as some men turned their efforts to the poultry business. Alva R. Badger of Toledo developed and began manufacturing the St. Helens Incubator, which won several prizes in a Seattle exhibit in 1901. That same year the *Bee Nugget* reported that C. F.

White of Boistfort, west of Winlock, spent most of his time caring for 350 chickens, and they were beginning to pay their keep.[54] The real start of the poultry business in the Winlock area is credited in local lore to John Marcotte, a Northern Pacific agent who moved to Winlock about 1910. In order to increase his commission based on the volume of freight shipped out of Winlock, Marcotte urged area farmers to sell him eggs, hogs, and veal, which he then resold to logging camps in neighboring Gray's Harbor county. In 1913 he formed the Cowlitz Produce Company to handle his growing business. It made $5,000 its first year, and by 1922 was taking in $2 million worth of produce, much of it in eggs.[55]

By the early 1920s, a mere 350 chickens would have seemed like a paltry number indeed. In 1921 an estimated 300,000 baby chicks were shipped into Winlock for the area's poultry farmers. Jacob Erving and John Annonen—who had the largest local flock with over 5,000 laying hens—built incubators in 1921 to try to fill some of the area's demand. Both efforts were highly successful. The Ervings lost most of their poultry operation to a fire in April 1923, but this was only a temporary setback. By June, the family was installing a new incubator with a capacity of 40,000 eggs, featuring automatic temperature control and egg turners, and planned to build a new flock of 6,000 first class breeder hens to keep it stocked.[56]

Throughout the 1910s, poultry farmers around Winlock were apparently content to market their eggs through Marcotte's Cowlitz Produce. Then in March 1920, 50 men met at the Finnish hall southwest of Winlock to form an association of egg and poultry raisers. The meeting was conducted in English and translated into Finnish, as many area chicken raisers, including the Ervings and Annonens, were members of the Finnish community. By the end of the summer, the cooperative had 175 members, represented 100,000 laying hens, and had affiliated with the Washington Cooperative Egg and Poultry Association.[57] The WCEPA rivaled the UDA in size and scope. It was formed in 1917, and in 1920, the year Winlock affiliated, it handled 2.5 million dozen eggs, paying the producers $1.4 million. By 1925 it handled over 11 million dozen eggs—about one third of statewide production—and paid its farmers $3.5 million. The WCEPA was more centrally organized than the UDA: members belonged to the central organization, not their local affiliate, and the Association directed the local units rather than vice versa. To assure representation, no county was allowed more than two of the organization's sixteen trustees.[58]

For several years the explosive growth of the egg and poultry industry allowed both Cowlitz Produce and the Winlock branch of the WCEPA to prosper and expand. In 1922 the *Lewis County Advocate* (renamed and now politically moderate, but descendant of the *People's Advocate*) reported that Lewis county was the number two egg producer in the country, after Petaluma, California.[59] Winlock eggs served markets as far away as London and Liverpool, with over half going to New York. The Extension Service reported that Washington eggs received higher prices than locally raised eggs in the East because of their consistent high quality and careful grading. On one weekend in January 1923, Winlock's Washington Coop sent 22,560 dozen eggs to New York, and Cowlitz Produce sent another 15,840. A few years earlier, Winlock area farmers had sold eggs and chickens primarily in western Washington markets. Now Winlock was the central processing point for the region, handling eggs from surrounding counties and even Oregon.[60] Both Cowlitz Produce and the Washington Coop built large plants to handle the volume. The Cowlitz Produce plant featured a new process called guaranizing that sealed the eggs and prevented spoilage. It employed forty people in round-the-clock shifts. The Washington Coop built a new three-story plant in 1923. That spring, the *Winlock News* excitedly reported that Cowlitz Produce had received a new order for sixty train cars of eggs for the New York market to be delivered over the next two months. The editor envisioned even larger orders in the future, predicting continued rapid growth in production and demand:

> But why stop at carloads. . . . In the big fruit section east of the mountains apples are handled in trainload shipments. Why not an "Egg Special" from Winlock to New York. It is not the least bit out of reason to suppose that such will happen soon. Three years ago we shipped a carload a month and referred to it with a great deal of pride. Right now they are going out at the rate of a carload a day. With another guaranizing plant or two we can send a trainload weekly and not get the least bit excited about it.[61]

The farming community around Winlock celebrated their new prominence in the egg and poultry world with a homecoming weekend in June 1923. Members of the Washington Coop and the Cougar Flat, St.

Urban, and Hope Granges provided food and entertainment. The *Winlock News* reporter estimated five hundred people attended and consumed gallons of strawberries and ice cream and "oodles" of cake and coffee.[62] The poultry industry continued to thrive for the next several decades, although Marcotte sold out in the mid-1920s. Winlock began calling itself the "Egg and Poultry Capital of the World." The town continues to host an annual "Egg Day Celebration" in June, complete with an Egg Day Queen. An enormous sculpture of an egg still stands next to the railroad tracks adjacent to downtown Winlock. Today's poultry farmers, however, have no equivalent of Darigold. Some farm families in the vicinity still raise poultry commercially, but under contract to the large distribution companies which supply the feed, antibiotics, and other components of the modern poultry business.[63]

If one were to read only government reports, one might well receive the impression that the Extension Service and its system of education, demonstration, and county agents was largely responsible for the success of the dairy and poultry industries in western Washington. Cooperative enterprises and stock improvements were among the chief priorities of the Extension Service. County agents and state Extension directors reported on the progress of agricultural improvements and cooperative organizations, whether or not they had anything to do with them, and followed the natural inclination to put their own work in the best light. The frustrations regularly expressed by a succession of Lewis county agents, however, make clear the marginal status of the county agent, an impression reinforced by reading the local press, which was more concerned with promoting the efforts of local citizens than those of outside professionals. Lewis county farmers wanted to improve their farming techniques. They read literature, attended lectures, visited each other's farms, let the county agent help them promote their own organizations, and even occasionally made the fifty-mile trek up to the Experimental Station at Puyallup once automobiles came into widespread use. They were not, however, willing to follow any new scheme proposed by "experts" that did not coincide with their own expertise as professional farmers, and they were unwilling to relinquish control of their social and economic lives by abandoning their own organizations in favor of new ones proposed by a county agent.

The often confusing, multilayered administration of the Extension Service and the county agent programs reflects the variety of sometimes conflicting interests that promoted them. The Smith-Lever Act of 1914 culminated several decades of national, state, and local attention to agricultural science and education by establishing a federally subsidized system of county Extension agents. The federal government, the state agricultural colleges, the counties, and local farmer organizations all contributed to the county agents' salaries, and all had a hand in directing their work. The Country Life Movement of the early twentieth century was part of the impetus behind the Smith-Lever Act. The Country Life Movement did not center around any particular organization, but around the desire of urban professionals concerned with agriculture to improve rural life. Faculty of state agricultural colleges, government bureaucrats, leaders of the emerging agribusinesses, and ministers in denominations with declining rural enrollments were among those who sought to make rural life more attractive, healthful, stimulating, and profitable, so that in the rush to urbanization America's traditional base would not be lost and cheap, plentiful food would continue to be provided for the cities. The variety of farmers' self-help and political organizations that flourished from the time of the founding of the Grange in the late 1860s to the Nonpartisan League and Farmers' Union of the 1910s, also contributed substantially to Progressive Era interest in farm life. These independent and politically active organizations were perceived as dangerous by many in the Country Life Movement, who sought to channel farmer concerns into more controllable directions.[64]

In Washington state, faculty of the State Agricultural College at Pullman began giving Farmers' Institutes around the state in 1893. The 1903 state legislature ordered annual Institutes be held for each county. The legislature established the Bureau of Farm Development to oversee the county agent program one year before the passage of Smith-Lever, and authorized use of Smith-Lever funds in 1915. At that time fourteen counties in the state, not including Lewis, employed county agents.[65] The question of hiring a county agent was discussed in Lewis county for several years. In 1913 and 1914 a number of letters on the subject were written to the *Chehalis Bee*. One author, calling himself a "struggling farmer," said he would appreciate receiving the advice of an expert. Another Cowlitz Prairie farmer summed up the opposite viewpoint by insisting he did not need "another useless professional man" telling him how to farm, adding that if county businessmen really wanted to help farmers they should buy more local produce rather than agitating for a county agent. The Pomona Grange took a similar position against the program, maintaining public money would be better spent improving farm-to-market roads. The Chehalis business community, on the other hand, was solidly behind the Extension Service and the county agent programs. In late 1913, the Chehalis Citizens' Club organized a Farmers' Institute at which resolutions in favor of the Smith-Lever Bill and hiring a county agent were passed, and a committee was appointed to meet with the county commissioners on the question.[66] Lewis county finally hired its first agent in 1917, a year in which twelve new Washington counties signed onto the program, induced, according to the state program director, by the war effort and increased support money from the federal government.[67]

According to his annual report, Lewis county agent A. B. Nystrom's activities during his first six months on the job included: writing seven newspaper articles; organizing seven boys' stock clubs with a total of sixty boys; helping one farmer build a silo; instructing six farmers (including one woman) in drainage; organizing an excursion of seventy-five farm men and women to the experimental station at Puyallup; holding nine Extension Schools which 1,125 people attended; and holding four canning demonstrations.[68] Over the next several years, Nystrom and his successors followed similar patterns of activity, and also spent considerable amounts of time trying to organize cow-testing associations and a county Farm Bureau—with little success.

The Extension Service generally promoted an ideal of sex-segregated work, with the farm husband as the businessman-farmer and the wife occupied with domestic concerns. In 1918, Mamie Nystrom worked alongside her husband as temporary home demonstration agent. She conducted twenty-eight demonstrations of canning and cooking without wheat to support the war effort, sometimes in mixed-sex settings such as Grange meetings. Lewis county did not hire a permanent home agent when the Nystroms left the following year, however, and apparently the wives of subsequent county agents did not take up the work. Consequently, "women's work" that was sometimes emphasized in other regions, such as sewing and home beautification, seems to have been relatively neglected in Lewis county.[69]

The Farm Bureau was as hybrid an organization as the Extension Service. It grew up in the 1910s alongside the county agent system. Local Farm Bureaus usually provided some of the money for the county agent's salary and provided an organizational base for Extension Service activities. With its formal national structure adopted in 1919, the Farm Bureau also provided insurance for members and lobbied the government for farm-oriented legislation. Its most avid supporters were prosperous, business-oriented farmers, small-town businessmen, and agribusiness representatives. The national leaders supported highly conservative politics, and while they sometimes cooperated with the Grange in certain lobbying efforts, most farmers' organizations, including the Grange, regarded the Farm Bureau and its business and governmental links with a great deal of suspicion. Just as the leaders of the LPDA on the local level tried to distance themselves from the roots of the farmers' cooperative movement, the leaders of the national Farm Bureau sought to distance themselves from the long history of farmers' associations by stressing that they were the first organization to represent farmers on a truly statewide or national basis.[70]

As frequently happened, the Lewis county agent became the chief Farm Bureau organizer once he was hired, even though the Bureau was supposed to be an independent, nongovernmental association. Nystrom, in reporting on his efforts in that direction in 1919, remarked on the importance of a Farm Bureau that would allow the farmers themselves to determine the kinds of help they needed from the county agent, as well as allowing him to conduct his work with the greatest efficiency: "I consider the Farm Bureau absolutely essential to efficient County Agent work. It makes possible the planning of the program of work by the farmers themselves. It makes for leadership among the farmers and certainly extends the work of the Agent into all parts of the County. When working with individuals alone the county Agent can serve but a few. Through the Farm Bureau he works with groups and his services therefore are of benefit to a larger number."[71]

Nystrom seemed optimistic that a Farm Bureau could be organized in Lewis county, but his successor, A. T. Flagg, acknowledged discouragement in his annual report for 1920. Nystrom had divided the county into sixteen districts and organized community committees for the Farm Bureau in eight of them by the time Flagg took over. At the end of his first year Flagg reported, "It is safe to say that not one community committee has worked with a definite knowledge of what it is to do." By 1922, a new agent, R. G. Fowler, stated there was only a "skeleton of a Farm Bureau organization" in Lewis county, and while some of the "strongest" farmers were behind it, it was "not receiving the support of the balance of the farming population."[72]

The fact that Lewis county farmers were already well organized, with twenty-two active Granges in 1920, apparently seemed but a hindrance to the county agents. In fact, between 1917 and 1920, while Nystrom and Flagg were struggling to build a Farm Bureau, eight new subordinate Granges organized in the county.[73] The state leader of the county agent program, R. B. Coglon, recognized the strength of the Grange and was much more sympathetic to it—and to the Farmers' Union which had adherents east of the mountains —than were the agents he directed. Coglon reported in 1918, "Lewis County is highly organized by the Grange. The communities are very old for this state and well defined. Most Granges of the county have excellent halls and are active in the interests of agriculture and social welfare."[74] Coglon reported that most of his agents throughout the state were reluctant to work with established farmers' groups, and he attributed this to the instinct for self-preservation. In most counties, the Grange or Farmers' Union was strong enough to have the office of county agent eliminated if the members chose. In 1916, Coglon wrote:

So far as I have been able to determine the Agent work in the State had been carried on without reference to local or county farmers' organizations. A number of the Agents availed themselves of opportunities to discuss farm topics at Grange or Farmers' Union meetings. Two in particular, according to their reports, had made use of the school system in their counties as a means of conducting their work. In no county, however, did I find that a systematic plan had been developed for building up local or county organizations of farmers for the purpose of providing ways by which the farmers themselves could solve their own problems, or for taking the farmers' organizations already existing in their counties, into their plans for agricultural development. The impression seemed to prevail among them that farmers' associations were dangerous bodies to deal with and were likely to be too effective in bringing about discontinuance of the office of Agricultural Agent should they become hostile or unmanageable for any cause.[75]

Many farm families were already in an organization dedicated to improving their social and economic lives, and one to which only farmers or certain other rural workers could belong. Most had no interest in joining another which purported to do the same thing, but which business leaders and academicians assured them, with some arrogance, could do it better because "professionals" and "concerned businessmen" were involved. Farmers questioned just whose interests would be served, not only by the Farm Bureau, but by the whole system of scientific farming proposed by the Extension Service. "Overproduction" and monopolistic distribution systems had long been the bane of farmers who suffered from widely fluctuating and frequently low prices. Increasing production without also controlling the distribution apparatus seemed risky indeed. Lewis county agent Fowler reported in 1922, "Many farmers seem to be of the opinion that we are producing too much, thereby keeping prices on farm products too low."[76] Thus farmers were highly selective in what they were willing to take from the Extension Service and county agents. The "conservatism" of farmers that so frustrated county agents and other observers reflected the conscious choice of politically active farmers seeking to control their economic destiny.

Lewis county farmers also chose the view of farm women advocated by the Grange over that of the Extension Service and Farm Bureau. Farm women, too, were members of the producers' alliance, as different from middle-class housewives as their Grange brothers were from town businessmen. In some ways, specialization and cooperative marketing shifted the division of labor within the family and reinforced the public image of the farmer as male. Making butter and raising chickens and eggs for family consumption and the market had traditionally been part of women's responsibilities, but as the dairy and poultry businesses became major income producers, farm men began taking control of the process. Women might well have continued to perform much of the labor, but it was the men who were considered to be the chicken ranchers in Lewis county newspapers and other accounts, and men who joined and managed the dairy and poultry cooperatives.[77] Women's work in agricultural production had always been obscured to some extent, but it is completely invisible when viewed through these organizations.

At the same time, rural men and women shared an ethic that valued hard work and recognized farming as a family way of life to which all contributed, not simply a job that a man might go to during the day.

This ethic remained in place through the 1920s and continues to be evident on family farms in the 1990s. The Grange explicitly recognized and celebrated this ethic by insisting on the inclusion of women and youth. The Extension Service, on the other hand, took the invisibility of women in farm work at face value, separating (male) scientific farming from (female) domestic economy, without recognizing the complexity that existed below the surface.

The invisibility of female agricultural workers was part of a society-wide devaluing of women's contributions. It preceded the commercialization of dairying and poultry and the success of the large cooperatives, rather than resulting from those developments. The male dominance of the cooperatives in the 1910s and 1920s grew out of the culture and the gendered division of labor of the farm. Men had customarily dealt with the public aspects of family finances and marketing. Men served as trustees and financial officers of all local organizations, even those of mixed sex. Apparently few if any questioned that men would run the cooperatives as well.

The very language of local people contributed to women's invisibility in agricultural production. Women signing Grange roles usually followed standard conventions and listed their occupation as housekeeper or farm wife. At the same time, local people clearly recognized that women contributed to farm family prosperity. When the Alpha Grange staged a debate on the perennially popular question, "that the farmer's wife does more work on the farm than the farmer does," the affirmative side won.[78] Even with the male monopoly over the cooperatives, women's expertise in poultry raising and dairying continued to be recognized in the twentieth century. A Cougar Flat Grange committee on dairy and poultry feed in 1921 included both men and women. Mrs. George Shoup of Puyallup was a frequent speaker to the "poultrymen" of the Washington Cooperative in Winlock in the early 1920s. In 1922, Mrs. Adda Howie, a cattle breeder from Milwaukee, attracted the largest crowd ever to a meeting of the Lewis County Jersey Breeders' Association—an organization which may have been mixed-sex, but had exclusively male leadership and quite possibly membership. According to the *Bee Nugget*, Howie was something of an eccentric who hung lace curtains in her barn, but she knew her subject. The paper praised her, saying, "Mrs Howie knows cows and talks most sensibly about them, making the point forcefully in her address that breeders should strive for straight, uniform cattle of constitution, capacity and udder development."[79]

As agricultural laborers in the hops and berry fields, rural women and their children continued to have a readily available way to earn cash. They earned far less than the men who worked in logging camps—a difference attributable in part to gender (and race, in the case of migrants), although certainly skill and danger were also factors. No doubt women and children did much of the picking because men's labor was viewed as more valuable on the home farm. Nonetheless, the money they earned was clearly their separate contribution to the family income, just as egg and butter money had been. Farms also continued to raise subsistence crops for family and home-stock consumption, with dairy farms keeping chickens and raising potatoes, and poultry farms keeping a cow or two.[80] In both time periods, men's work was more often publicly acknowledged and was evidently valued more highly than women's, but men and women nonetheless shared work on the farm, with certain fairly flexible areas of specialization.

In this setting, the Grange and the ethic it endorsed mitigated the potential economic power of men over women, although it certainly did not prevent abuses. A drunken, abusive, or simply lazy husband or father could do a lot more damage to his family with the economic and other powers the law and society gave him than a woman could do to hers, as local newspaper accounts of domestic violence and evictions amply document.[81] Some recent studies of American farm women have emphasized their hard and unremitting labor and vulnerability to abuse. In these accounts, isolation and poverty are major contributors to women's hard lives. Lewis county farms were, however, generally close to neighbors and relatively prosperous. Kin relationships and ethnic communities were strong, and community associations were dense, even in the countryside. The Grange ritual and the easy give-and-take at meetings fostered respect for women's separate work and their intellectual capacities. These factors helped mitigate the hardness and dangers of farm life for many Lewis county women.[82]

In the early twentieth century, Lewis county farm families gained a measure of the control they had long sought over their economic well-being through their cooperatives. They could not control the ups and downs of business cycles, but they were no longer entirely at the mercy of distant profit-seeking distributors. Many Lewis county farmers were able to do well through the generally devastating agricultural depression of the early 1920s as a result. In building the cooperatives, farmers were acting on goals that had been articulated by the Farmers' Alliance and Populist party two decades earlier. The LPDA and even the smaller, locally organized cooperatives were, on the other hand, strictly economic concerns and were separated from the heart of rural communities. The cooperatives reinforced the sense of shared interests between town and country dwellers and maintained a careful distance from the Grange. Thus they did not provide the kind of political training in radicalism that the Texas Farmers' Alliance Exchange did in Goodwyn's analysis. At the same time, Lewis county farmers remained committed to the Grange and resisted efforts by businessmen's associations and county agents to lure them away from the producers' alliance and their concept of participatory democracy. As the political spectrum narrowed and tolerance declined with the entry of the United States into the World War, that resistance took considerable courage.

$$\mathcal{C}\!\!\!\!\!\!\text{ʌ } 7 \text{ ʌ}\mathcal{D}$$

A Community in Conflict:
The End of Tolerance

While the Grange and agricultural cooperatives became fixtures of rural neighborhoods, the local towns were also changing. The booming timber industry and new railroad construction attracted many new workers. Napavine more than doubled in population between 1900 and 1910, and Winlock and Little Falls nearly did. The population of the towns waned during the 1910s as the timber industry moved on to stands of virgin timber, but some of the mills remained, and with agriculture prospering, most businesses remained strong. Class lines in the towns hardened with their increasing maturity. Timber and railroad companies brought in single male immigrants after the turn of the century who remained clearly outside community boundaries. At the same time, "American" and Americanized residents of the towns increasingly gained the wealth and leisure to adopt a more genteel lifestyle that further separated them from both the timber laborers and hardworking farm families. The Masons and Oddfellows continued to include farmers and the more "respectable" mill workers among their members, but new organizations were also formed exclusively of townspeople with the goal of promoting development of their towns along "desirable" lines.

Long established and financially secure farmers close to the towns shared much in common with business leaders, and the cooperatives fostered their connections. Nonetheless, the political stance of the Grange and of the town business groups increasingly diverged, and when the United States entered World War I, the political fluidity that had characterized the early 1910s came to an abrupt end. Up until that time, activist farmers had been viewed by their more conservative neighbors and townsfolk primarily as misguided fools, not as a threat to American democracy. But the accelerating conflict between timber workers and mill owners and the sharp crackdown on all forms of dissent by the federal and state governments during and after the war created a far more violent atmosphere, with lines of conflict clearly drawn and a very narrow definition of patriotism legitimated. The dramatic spectacle of the Russian Revolution was brought uncomfortably close to home for many with the Seattle General Strike and the bloody fight between Wobblies and American Legionnaires in Centralia in 1919. People who had never liked the political leanings of Lewis county farmers began to feel justified in moving to stamp out what was now clearly defined as radicalism, even if it took violence to do so.

The booming timber industry (see Figure 9) needed more workers by the early twentieth century than natural migration to the area provided. Along with the farm men and other well-established groups that worked in logging camps and mills, the timber and railroad companies now also brought in groups of male immigrants from southern and eastern Europe and Asia (see Figure 10). Many of these workers lived in boarding houses which ranged from small family-run affairs to large company houses with twenty or more men from the same country, working for the same employer. For example, at the time of the 1910 census, a young Hungarian couple in Little Falls housed themselves and their two small children, his brother, and six other Hungarian men, all of whom worked at the clay factory. Nearly next door lived fifteen men, all laborers at a sawmill who had arrived from India in 1906. Another Little Falls boardinghouse was home to forty-three Bulgarian men who had immigrated between 1906 and 1909 and worked as laborers for the Northern Pacific Railroad. Nearby, a native-born couple with five children housed nine American and British railroad workers. Winlock's numerous boardinghouses included one run by a divorced woman with the help of her mother and sister, housing five American and northern European skilled workers; another with thirty-six Italian railroad laborers; one with eight Greek sawmill employees; and one with sixteen Finnish logging-camp workers.[1]

Of the male boarders in the heavily industrialized precincts of

Figure 9. Summerville Brothers mill, Napavine, Washington, before it burned in 1909. Photograph courtesy of Lewis County Historical Museum, Chehalis, Wash.

Figure 10. Summerville mill employees, Napavine, Washington, ca. 1900. Proprietors are seated near center. Asian employees are in the front row. Photograph courtesy of Lewis County Historical Museum, Chehalis, Wash.

Napavine, Ainslie, Prescott, Veness, Little Falls, and Winlock in 1910, sixty percent were immigrants, compared to thirty-seven percent of the total adult population for those precincts. These men varied considerably in life circumstances. Many were young and single—on average, they were four years younger than the other adults in the area. Some were married, but few were accompanied by their wives. They presumably intended to return to their homelands after saving enough money. These men did not join existing community associations and probably would have been rejected had they made the attempt. They were viewed as temporary residents.

No overt acts of hostility against foreign workers in the study area are recorded, but in nearby Centralia and Chehalis, Greek and Italian workers were run out of town in 1910 and 1911. The *Bee Nugget* reported in June of 1910 that a Centralia man had been accused of "ruining" a fourteen-year-old girl. The man, whose last name was Stewart, was presumably not Italian himself, but he boarded Italians. Stewart was in jail, and an angry mob had driven the Italians out of town. The article does not connect the Italians with the rape, but apparently some perception of degraded morals linked the innocent boarders with the alleged criminal in the eyes of the mob.[2] The following summer, Chehalis residents held a series of meetings to organize an anti–Greek and Italian association. The *Bee Nugget* reported six hundred people attended one meeting. Speakers denounced the hiring of Greeks and Italians while "white men" were out of work, and asserted that the foreigners lived like cattle in crowded boardinghouses, creating a bad impression for more "desirable" newcomers. By the end of July, most of the Greek and Italian workers had left town—some without paying their grocer bills, and a Greek pool hall and bakery had closed.[3]

Despite the apparent animosity of many native-born residents, the mills continued to hire foreign crews. The financial advantages gained by paying foreign workers less than native-born men outweighed any negative effects. In 1921 representatives from fifteen Lewis county mills met and agreed to lower their base pay by $1.00 to $3.20 a day. But Japanese crews would be paid $2.80, and other foreign laborers only $2.50.[4]

In the meantime, most German and other northern European immigrants continued to enjoy the positive attitudes that had prevailed in the 1890s. The 1911 newspaper issue that reported the forced departure of the Greeks and Italians from Chehalis carried another article on the same page that described the great success of the German picnic held the week before. "Their open meetings are popular events," the *Bee Nugget* said, "for they are good entertainers, and like to see everyone have a good time."[5] Like the Germans, the Finns were viewed by most native-born Americans as adopting "American" values and living compatibly with their neighbors, although they did form exclusive foreign-language organizations. The Finns came in family groups, bought land, worked hard, went to church, and formed voluntary associations for recreation and cooperative endeavors. By the early 1920s, Finns who had been raised in America were joining English-speaking groups. Winlock's Oddfellows lodge had two Finnish officers in 1923, Matt Torkko and Erik Bay. Winlock's baseball and basketball teams, which competed hotly with teams from nearby towns, included many young Finnish men. The West-Side Neighbors Club of Winlock held monthly parties in the early 1920s, including people from different generations and with Finnish, German, and British last names. Obituaries in the *Winlock News* spoke highly of the hard work and community service of departed Finns.[6]

With the influx of new population and the economic shifts the region experienced, community associations remained strong, helping integrate "desirable" newcomers into social networks and community norms. The mix of associations in the first decades of the twentieth century was similar to what it had been in the 1890s. Churches remained numerous as did the fraternal lodges. "Leading" townspeople also organized single-sex clubs to boost their communities. A Toledo Improvement Club had existed in the mid-1890s, but apparently did not last long. A Toledo Commercial Club, composed largely of businessmen, is mentioned from time to time in county newspapers in the 1910s and early 1920s, discussing such issues as improving the volunteer fire department, protecting the Cowlitz bridge from floods, and outfitting a tourist park.[7] Toledo women organized the Toledo Civic Club and sponsored town beautification projects from the early 1910s into the 1920s. Winlock's Ladies' Civic Club paid for improvements to the town park, initiated a cleanup day, distributed Christmas boxes to the needy, and pro-

posed building a community center. Winlock men started several short-lived clubs early in the century before organizing the Commercial Club that lasted from 1921 into the 1940s. Vader and Napavine had similar clubs.[8]

One of the leading members of the Winlock Commercial Club
was John C. Lawrence, manager of the Washington Egg and Poultry
Cooperative. He and others sought to unite farmers and businessmen in common endeavors, to encourage settlement on still vacant
land, and to promote commercial enterprises such as canneries.
The *Winlock News*, in reporting on the reorganization of the Commercial Club in 1923, bemoaned the lack of "co-operation and
pull-together spirit," claiming "the country people and the people
in town have drifted farther apart." The *News* editor hoped the
club could help overcome the problem.[9] There is little evidence,
however, that country people took any part in the commercial and
civic clubs, despite the high profile of people like Lawrence. All of
the members of these clubs in the 1910s and 1920s who could be
traced in the Polk Directories of the time were town dwellers, and
most were clearly of the more prosperous business class, including
business owners, bank managers, physicians, newspaper editors,
and real estate agents. A Winlock ladies' bridge and literary club of
the early 1920s, the Olequa Club, was similar in membership if not
in purpose. A few of the women were wives of skilled workers, but
most were either professional women or businesswomen themselves, or were married to businessmen. All of the members of
these clubs whose names were pulled from newspaper reports were
also members of at least one, and in most cases several, other organizations.[10]

The fraternal lodges during the early twentieth century retained
a fair amount of diversity in occupation and ethnic background of
members. At the same time, lodge members shared longer-than-
average residence in the community and were wealthier and more
likely to be of native parentage than other residents of the study
area. Male lodge members were also considerably more likely than
average to be in business or a profession. About one-third of Masons
and Oddfellows were farmers, but no Woodmen were, and many
fewer lodge members were laborers than among the general population. Women lodge members were also about twice as likely to
be in business or a profession than average.[11]

In terms of wealth, occupation, and nativity, lodge members

Table 12. Occupations of men in western precincts of the study area by precinct and organization, 1910

Precinct/town	Number of men	Occupation					
		Farmer	Timber worker	Railroad laborer	Other labor	Business/ professional person	Student/ retired
All men	2324	25.5%	31.2%	12.7%	14.9%	9.4%	4.2%
Farming precincts							
Cowlitz	132	78.0	1.5	0	0	2.3	7.6
Drews Prairie	38	84.2	2.6	2.6	0	0	2.6
Eden Prairie	64	70.3	7.8	0	9.4	0	10.9
Economically mixed precincts							
Ainslie	90	30.0	55.6	3.3	4.4	4.4	2.2
Napavine	592	20.6	37.0	24.0	13.0	5.4	1.2
Prescott	149	41.6	20.8	2.0	25.5	4.0	2.0
Salmon Creek	158	49.4	17.7	1.3	8.9	3.8	8.2
Stillwater	130	50.8	19.2	15.4	6.2	1.5	3.6
Veness	98	33.7	46.9	0	15.3	0	1.0
Incorporated towns							
Little Falls	261	0.8	30.7	26.1	28.0	11.9	2.3
Toledo	135	3.0	18.5	0	30.4	29.6	17.0
Winlock	477	3.8	44.4	11.5	14.9	19.9	4.0
Organization							
Masons	87	32.2	17.2	0	11.5	27.6	11.5
Oddfellows	14	35.7	7.1	0	7.1	42.9	7.1
Eastern Star	46	28.3	19.6	0	10.9	32.6	8.7
Rebekahs	23	43.5	17.4	0	21.7	17.4	0
Woodmen	10	0	20.0	0	0	70.0	10.0
Grange	251	74.1	9.2	2.8	5.6	3.2	5.2

tended toward the highest-status groups. Nonetheless, the lodges did not become bastions of elitism in the decade after 1900. A significant portion of lodge members continued to be laboring people, and there were as many farmers as businessmen in most lodges. Acceptability for lodge membership did not so much depend on occupation—lumber barons and mill laborers, town merchants and farmers belonged (see Table 12 and 13). Perceived connection to the community and conformance with community norms were much more important, just as they had been in the 1890s. The Japanese, Bulgarian, and Italian men who lived and worked in the area, contributing to the overall prosperity of the region, were never invited to join the lodges, we can safely assume. Living in large groups,

Table 13. Occupations of women in western precincts of the study area by precinct and organization, 1910

		Occupation				
Precinct/town	Number of women	Housekeeper	Farmer	Laborer	Business/ professional person	Student/ retired
All women	1446	81.1%	2.0%	6.6%	6.4%	5.6%
Farming precincts						
Cowlitz	103	87.4	2.9	0	3.9	5.8
Drews Prairie	38	92.1	2.6	2.6	0	2.6
Eden Prairie	51	82.4	0	0	7.8	9.8
Economically mixed precincts						
Ainslie	72	83.3	2.8	6.9	0	4.2
Napavine	279	79.2	4.7	6.1	6.5	3.2
Prescott	94	86.2	2.1	6.4	2.1	3.2
Salmon Creek	99	80.8	0	7.1	3.0	9.1
Stillwater	62	87.1	3.2	3.2	3.2	3.2
Veness	78	88.5	5.1	2.6	1.3	2.6
Incorporated towns						
Little Falls	142	78.2	0	7.0	11.3	2.8
Toledo	111	65.8	0	9.0	8.1	16.2
Winlock	317	80.8	0.6	11.0	10.7	6.0
Organization						
Eastern Star	74	79.7	0	6.8	12.2	1.3
Rebekahs	48	70.8	0	4.2	12.5	12.5
Woodmen	28	89.3	0	3.6	7.1	0
Grange	110	88.2	4.5	0.9	5.5	5.4

their wives and children still in the homeland, these men were viewed neither as respectable nor as desirable settlers. Fraternal lodges might well have inculcated middle-class values as some historians have claimed, but only in those people already deemed suitable to join.[12]

While most people joined one association and devoted their available time to it, some people belonged to a number of organizations, sometimes changing with different stages of their lives. Myrtle Fletcher joined Hope Grange in 1909 as a twenty-seven-year-old housewife, married to a surveyor and living in the rural neighborhood of Evaline. In the early 1910s, she and her husband moved to Winlock. There she joined the Eastern Star, serving as Worthy Matron in 1916 and 1917. In the early 1920s she also belonged to the Neighbors of Woodcraft, the Winlock P.T.A., and was an officer in the Winlock Baptist church. Mattie Morton was also a former mem-

ber of Hope Grange, who as the wife of a real estate agent in the
1920s participated in the Neighbors of Woodcraft, the Winlock
P.T.A., the Rebekahs, and the Methodist church. Peter Severson, a
Toledo shoemaker born in Norway, joined the Toledo Masons in
1909 at the age of forty-six, followed shortly by his sons, Arthur and
Oscar. He was a charter member of that town's Eastern Star lodge
in 1919, along with his wife, Ovedia, and daughter, Nina, and was
also active in the Modern Woodmen. In 1922 he won election as a
town councilman. The other Seversons, but not Peter, were all active
in the Eadonia Methodist church as well. Matt Torkko joined the
Finnish Kaleva lodge in 1915. In the early 1920s, Torkko participated in the Winlock's West-Side Neighbors Club, the Oddfellows,
the Winlock town baseball team, and a purebred sires club.[13]

Members of Winlock's Neighbors of Woodcraft and Toledo's
Woodmen and Royal Neighbors were especially likely to belong to
other groups. Only one person identified in these organizations was
not also found in other associations. The Woodmen were also the
most urban of any group, with no members occupied as farmers.
Living in town, they had the most ready access to the various associations, and the women had the most leisure. Most members of
other lodges apparently belonged to only one fraternal association.
At the same time, participating in two or even three was not uncommon and was not considered in any way improper or disloyal.

Only a small minority of Grange members, in contrast, participated in the lodges, although a handful could be found in each.
Just over ten percent of the 1910 sample of Grangers later joined
the LPDA. Probably a number of others belonged to the other dairy
and the poultry cooperatives for which membership lists are not
available. The cooperatives were very different kinds of organizations, however, with membership clearly a part of work life rather
than a voluntary leisure activity. LPDA members were even less likely
to be in a lodge or church, but over forty-five percent of them could
be found on Grange rolls.[14]

While churches did not allow dual memberships, that policy did
not prevent other kinds of mixing. All of the churches had some
kind of women's group, and some women participated in their
friends' or relatives' church groups as well as their own. In the early
1920s, Mollie Baldwin of Winlock was active in the Ladies' Aid Society of the Methodist church which she had joined in 1919, and in
the Baptist Ladies' Aid. She also belonged to the Eastern Star, the

Neighbors of Woodcraft, and served as president of the Winlock Civic Club. Emma Hoffmann, an officer in the St. Urban Grange in the early 1920s, participated in both the St. Urban Catholic Church Altar Society and in the St. Peter's Lutheran Church Ladies' Aid. Freddie Yansen was active in the Ladies' Aids of both St. Peters' and Winlock Methodist, as well as the Eastern Star, the West-Side Neighbors Club, the Olequa Club, the Winlock P.T.A., and the Lewis County Minute Women. People occasionally changed church membership, which might account for some of the joint participation. The women's groups were largely social in nature. Ties of friendship might remain even if some other spiritual reason or conflict or a change in life circumstances precipitated a move.[15]

Those who could be identified as participating in several organizations were on average three to four years older and had been married three to five years longer than those found in only one association. Multiple group members were also consistently less likely to be laborers and more likely to be businesspeople or housewives than the less active group, and were more likely to hold both real and personal taxable property. Among Grangers, the active group's average real property assessments were somewhat higher than the less active group's. In nativity, there was no consistent variation among the more and less active groups. Once they had joined one association, immigrants were no less likely than the native-born to join another.

People who were active in several associations needed to have some leisure in which to do so. Their relative wealth clearly helped purchase that leisure. For women, the number and ages of their children were also major factors in their freedom to be involved in community life. Nonetheless, in most of the organizations analyzed here the majority of women members were married, had children, and were younger than forty-five. Of these younger women, lodge members had on average two living children. Grange and church women were more fertile, with on average four children born and three surviving. The women over age forty-five in the various organizations had given birth on average to six children. Many of them had lost at least one of their children.[16]

Across the United States, rural fertility rates were significantly higher than urban ones in this period, although both decreased steadily through the nineteenth century. White American women surviving to menopause in 1910 bore on average 3.4 children.[17]

Lewis county women in the study area over age forty-five, whether native or foreign born, had given birth on average to six children, according to the 1910 census. The size of the towns studied here hardly qualifies them as "urban," and the women living in town shared in the generally high fertility. While total fertility rates did vary somewhat by precinct, from a low of five to a high of eight for Drews Prairie, the overall average for all women living on farms was the same as for nonfarm women. Among the younger women, however, farm women and immigrants had more children than native-born women living in town. Married, native-born, nonfarm women under age forty-five averaged three children, while farm women averaged four.

The meaning of these fertility patterns for women's community participation is not completely clear. As we would expect, lodge women, who tended to be the wives of town businessmen, in general had fewer children to care for than the rural Grange women and the more diverse church women. At the same time, having children was the norm and did not preclude community activity. Organizational minutes show women resigning officers' positions with some regularity, only to reappear as active participants a short time later.[18] We can assume that at least some of these temporary withdrawals were because of pregnancy—but they were not the prolonged absences that childcare would require. Women active in several organizations did not have fewer children than those in just one. In fact, because the more active group tended to be older and longer married, they frequently had more children. The Ladies' Aid societies, Grange Auxiliaries, and less formal groups like the West-Side Neighbors' Club were structured specifically to allow women to bring their young children along. The Winlock Eastern Star had a sewing circle in addition to their regular meetings which made attendance by young mothers easier.[19] Children certainly accompanied their parents to church, and often evidently to Grange meetings. Town women probably had a greater variety of childcare options immediately available, such as family members, neighbors, and in a few cases, servants.

What emerges from this study of organizational membership is neither clear-cut lines of class or town-country differences, nor a complete blurring of such lines. Farmers, business people, and well-established laborers mixed in the lodges and churches in a setting that presumed their fundamental equality—at least in principle. A small but active core of individuals belonged to a number of associations, expanding circles of friendship and assuring the continuity of those commonly accepted values and patterns of activity that were so evident in the 1890s. The Granges, lodges, churches, and civic clubs had much in common: all emphasized fraternity, family, community service, and social amusements. At the same time, most people participated in only one social organization, and the differences could be pronounced. The political education received by Grangers in their explicitly class-based organization, and the activist agenda that resulted, were a far cry from the officially nonpolitical stance of the lodges located in politically conservative towns. Few Grangers joined other associations to dilute their agrarian bent. The commercial and civic clubs in the towns also had a strong class orientation—although the members probably would have denied it—and in a very different direction. The members of these clubs were active in a broad range of activities, and saw themselves as community leaders with the interests of the whole area at heart. But the town business people and the "producers" on the farms and in the mills frequently defined those interests quite differently.

In Lewis county, the heightened tensions of the postwar years appear to have grown more from the increasingly hard-line position of conservatives than from an increase in radicalism in the countryside. Even in the Northwest, Lewis county is best known for the so-called Centralia Massacre. Consequently, radicalism anywhere in the county is frequently associated with the Industrial Workers of the World, or Wobblies. The evidence gathered for this study, however, suggests little connection between the efforts of the IWW to organize the lumber industry and the consistent support of rural precincts for political third parties, despite the importance of lumber in the local economy and the fact that many farming families had at least one member in the mills or timber camps. Wobblies are rarely mentioned in any of the sources as either a problem or as allies prior to the big strike of 1917 that swept the northwest forests, and even then Grange minutes record no discussion of the IWW or the strike. The bulk of third-party voters appears to have come from the core of propertied farmers with long roots in the community and the newly arrived prospective farmers seeking the cash to build up their land. These were not the same people the IWW sought to organize. Nonetheless, the conflict between workers and owners in the lumber industry was a major factor in the polarization of the community in south-central Lewis county, as elsewhere in the region. The matter is therefore worth examining here.

The lumber industry throughout the Northwest was a highly volatile one, with intense competition between overly abundant mills, sharp swings in prices and demand, and high physical and economic risks. Among the consequences were low pay and deplorable conditions for workers, particularly in the logging camps. Logging camps were by nature temporary—once the trees were cut, workers moved on to virgin stands. Loggers worked long hours in all kinds of weather with great safety hazards. They then returned to crowded, poorly built, vermin-infested bunk houses with large quantities of bad food, and no showers, laundry facilities, or places to dry their clothing. They carried their own bedrolls from job to job. In between, loggers gathered in the cities and towns of the Northwest, spending their money and waiting for the next job.[20] The IWW located halls near the urban hangouts of the loggers and provided the men with a number of services as well as radical literature and a sense of family, thereby gradually building a loyal following. The AFL's International Union of Shingle Weavers, Sawmill Workers, and Woodsmen also attempted to organize loggers, but not until the labor shortage and price increases of the war created an opening did labor organization make much headway.

Conditions in the mills were not quite as bad. At least living conditions were more stable. In the study area, most workers lived in their own homes with their families, although some lived in large boarding houses, and the more isolated villages elsewhere in the region could be virtual company towns. The work was quite dangerous, surrounded by open saw blades and little in the way of safety equipment. Few who spent much time in the mills kept all their fingers, and more ghastly accidents were not uncommon. In May 1901, the *Bee Nugget* reported the third fatal accident in two weeks within twenty-five miles of Chehalis and concluded that timber work was more hazardous than serving in war. Work was also unstable. Mills closed frequently because of fires, bad weather, or unfavorable economic conditions. Both the AFL and the IWW tried organizing northwest mill workers, but were only successful in a few towns and made little effort or progress in Lewis county before the war.[21]

Mill owners, in the meantime, often built up comfortable fortunes despite sharp market swings, and fared somewhat better at collective efforts. A Southwest Washington Lumber Manufacturers' Association existed by 1900, representing many Lewis county mill owners. They set rates among themselves and also met with other regional associations from western Washington and Oregon to discuss rates and grading. In October 1912, most shingle mills in Washington shut down in a collective effort to raise prices.[22] Such local efforts at cooperation, however, did not succeed in stabilizing the industry. In the summer of 1917, the IWW took advantage of wartime conditions to declare a strike of loggers, demanding an eight-hour day; a minimum wage of sixty dollars a month with free board and clean bedding provided; a maximum of twelve men per bunk house, with reading lights, showers, and laundry facilities; and hiring at the union hall or job site, with transportation provided. Despite little in the way of standing organization, the strike succeeded in virtually closing down the industry in the Northwest. Mill owners quickly countered by organizing a Lumberman's Protective Association, pledging themselves to hold out against all union demands and establishing penalties for any lumberman who gave in. By fall, the loggers were out of money and began returning to work, but the strike simply took on new forms, with slowdowns and regular work stoppages after eight hours that continued to disrupt production.[23]

Confrontations between the Wobblies and mill owners had ended violently in the past. At least seven people had died in Everett in 1916, when IWW supporters came to assist striking shingle-mill workers and tangled with law enforcement officials and armed citizens opposed to the strike.[24] Such violence was prevented during the 1917 strike, when the federal government intervened in the name of the war effort. Government negotiators pressured lumbermen for concessions, well aware of the deplorable conditions in lumber camps as well as the need to increase production. At the same time, the federal and local governments launched an all-out attack on the IWW, which was opposed to the war. Across the country, Wobbly leaders were arrested under newly passed antisyndicalism laws which made simply belonging to the organization a crime. This effort was aided by vigilante groups of citizens who smashed up IWW halls in a number of cities.[25] Work slowdowns and disruptions continued in the logging camps nonetheless. In 1918 lumbermen finally gave in, and timber workers won the eight-hour day and the right to company-provided clean bedding, but not the right to be represented by their own union.

Lewis county remained relatively quiet during the timber strike. Some men walked off their jobs during the summer of 1917, but many mills kept running. The *Toledo Messenger* reported a strike in the remote mountain camp of Vance in eastern Lewis county in July. According to the account, the men walked out only to save their jobs after Wobblies threatened to burn the plant if they remained. The same issue reported that owners of the Stillwater mill near Vader had provided their workers with arms and ammunition to guard against the IWW.[26] In the meantime, mill owners remained united. Twenty-five lumbermen met in Centralia in February 1918, prior to the final strike resolution, and agreed not to give in to the eight-hour day.[27]

The IWW had never succeeded in establishing a base in the county—in part because of constant persecution. An IWW member was arrested in Chehalis for vagrancy in 1912 and held over night, then told to move on. "Hoboes" continued to be a problem in the town, prompting an addition to the police force in 1913. A much larger group was run out of Centralia in 1915, although they were apparently looking for food and work, not seeking to organize.[28] Wobblies arrived with more serious intent in 1917 with the strike in full swing. They rented a hall in Centralia to serve itinerant loggers. The organizers were evicted from their first quarters but found new ones. Then during a Red Cross parade in early April 1918, a group of Centralians raided the hall, burned the furniture and papers, beat the Wobblies, and ran them out of town. The *Toledo Messenger* called the Centralia raiders "patriotic citizens." An editorial in the *Chehalis Bee Nugget* defended the right of the citizens of Centralia to engage in "house cleaning." While generally frowning on mob action, the editor declared, "when the authorities fail to act in a manner which it is clearly their duty to pursue, we can hardly keep from winking the other eye or looking at the beautiful, fleecy clouds, if an outraged public acts occasionally in matters of this kind."[29]

A few weeks later the Chehalis city commissioners passed an ordinance giving the police authority to close up and dispose of any property used by the IWW or any other organization preaching "sabotage, violence or unlawful methods of terrorism." That June, a man was arrested in the logging town of Pe Ell for distributing IWW literature. He reputedly "bragged" of not having worked in twelve years, thus proving in the eyes of the *Bee Nugget* editor that such radicals were lazy as well as dangerous. The *Toledo Messenger* had earlier branded the IWW a "traitorous band which won't work or allow anyone else to work."[30]

The IWW was well accustomed to adversity, however, and once again opened a hall in Centralia in 1919. Members of the local Elks and the American Legion, newly formed by World War veterans, immediately formed a committee to take care of the problem once and for all. Rumors of another attack on the hall circulated. When the Wobblies discovered that the upcoming Armistice Day parade would pass the new IWW hall, they concluded that the parade would again be the occasion of an attack. After consulting with a lawyer, they decided on an armed defense. On Armistice Day several nervous men gathered in the IWW hall with their guns, while others hid in nearby vantage points. The Centralia American Legion was the last group in the parade. They fell well behind the rest of the marchers and stopped in front of the IWW hall, in order to tighten up their ranks, the leaders later testified. According to the press reports widely circulated after the event, the Wobblies proceeded to gun down in cold blood the unarmed Legionnaires. "4 Ex-Soldiers Murdered Tuesday, IWW shot into Armistice Parade," screamed *Bee Nugget* headlines. Much of the testimony taken from the Legionnaires themselves, however, suggests that shots were fired only after several men had broken ranks and stormed the hall. Four Legionnaires were killed that day, and one Wobbly, Wesley Everest, was lynched that night by a group of masked men who abducted him from the jail house. For weeks thereafter, posses rounded up suspected Wobblies. Eleven were tried for murder in the deaths of the Legionnaires. Eight were convicted of second-degree murder. No investigation was ever made of the lynching of Wesley Everest or the destruction of the IWW hall that followed the shootings.[31]

Emotions ran high following the event. In nearby Chehalis, the *Bee Nugget* declared, "Feeling . . . almost equalled the feeling in our sister city. More than ever the determination is expressed here that never shall the IWW get a foot hold in Chehalis."[32] In Seattle, two labor newspapers were shut down by the authorities for daring to suggest that a thorough investigation of the incident should be conducted. In towns throughout the West and in other parts of the country, news of the Legionnaires' deaths prompted retaliatory raids by angry citizens on IWW halls.[33] Still today, few in Lewis county are willing to discuss the event. The IWW did not give up, however, nor did the biased reporting of the "massacre" eliminate all Wobbly support. Over the next several years, a number of men were arrested in Centralia, Winlock, and other locations for distributing IWW literature or simply carrying IWW cards. In 1923, the *Winlock News* reported "a sporadic outbreak of wobblyism, attended with little violence," when forty to seventy-five men (depending on the witness) walked off the job at a Winlock mill to attend a May Day parade in Portland.[34] After the conviction of the Centralia Massacre defendants, the IWW and sympathetic supporters tried for years to win their release. The *Winlock News* reported in early 1922 that Wobbly "friends" were "flooding the district" soliciting funds to continue the campaign to free the convicted men. "Don't be fooled into forwarding money to those pirates," warned the paper. "Any funds contributed will go to enrich the fat pocketbooks of the higher ups instead of being used to get releases for anybody." Interestingly, although the paper referred to the prisoners as "murderers," it made no argument against contributing to their defense on the basis of their supposed guilt. The same issue lauded the "patriots" who contributed to a memorial monument for the slain Legionnaires. It was not until the 1930s, after several church organizations investigated the affair and began petitioning on their behalf, that the Centralia prisoners were finally freed.[35]

Persecution of the Wobblies was only one aspect of a broad crackdown on reformers during and after the war. Many of the people involved in the loosely knit reform coalitions of the Progressive period opposed both the war and United States involvement in it. To many Socialists, the war was a battle among capitalists, with working men used as cannon fodder. Others viewed it more simply as a distraction from more pressing social and economic problems. To those who had opposed the reform movement, on the other hand, wartime patriotism provided an ideal cover to question the loyalty of their political opponents. Any labor action could be deemed traitorous. While also warning employers to act reasonably, the *Toledo Messenger* declared in August 1917, "Every strike by labor, organized or otherwise, at this time is a blow struck for the Kaiser." The IWW was so dangerous in the eyes of the *Messenger*'s editor that he asserted, "it's high time that every man carrying an I.W.W. card be arrested and tried for treason in a time of war."[36] The IWW was one of the chief targets, but other organizations fell victim to the antiradical hysteria as well. One of these was the Washington State Grange, despite the Grange's repeated statements of patriotism and support for the war effort.[37]

The Grange's entanglement with the Nonpartisan League provided much of the immediate provocation for attacks. The Nonpartisan League (NPL) originated in North Dakota in 1915 and by 1918 was at the peak of its strength. The population of North Dakota was dominated by wheat farmers, but the grain elevators, mills, railroads, and grain exchanges on which the farmers depended were in the hands of businessmen in Minneapolis and other large midwestern cities, while a small group of Republican party loyalists controlled state politics. Former Socialist Arthur C. Townley set out to organize farmers to enter the political process. The Nonpartisan League was more centralized than the Farmers' Alliance had been, but followed a similar strategy of tapping into existing community organizations and establishing its own press to educate farmers. The NPL was able to capture the state Republican primaries, nominate and elect its own candidates, and thus win near control of state government in 1917 and 1919. The new government passed sweeping reform legislation, including provisions for a state bank, state-owned grain elevators and mills, hail insurance, and workman's compensation.[38]

Such far-reaching reforms were enough in themselves to earn the NPL the enmity of the old guard. When Townley declared the war effort irrelevant to farmers, criticized war profiteers and low wheat prices, and pressed on with reform, he was labeled unpatriotic and disloyal as well.[39] At the same time, League organizers spread their gospel to other states including Minnesota, where it was part of the origins of the Farmer-Labor party of that state, and Washington, where similar strategies of nonpartisanship had been practiced by the reform coalition for years. Leaders in the State Grange welcomed the organization. An NPL official spoke at the state meeting in 1917, followed by Kegley, who strongly urged the political involvement of all Grange members. By the spring of 1918, the NPL was active in many parts of the state, but the movement's reception was decidedly mixed. While a number of farmers evidently joined, vigilante violence was directed against many of the organizers.[40] One particularly nasty series of incidents took place in Winlock and Toledo in late April 1918, just three weeks after the first destruction of IWW headquarters in Centralia.

For several months, the *Bee Nugget* had been reporting on the organizing activities of the NPL in the county, warning residents of the League's unpatriotic reputation and hinting of ties to the IWW and pro-German forces. On April 26, A. Knutson, NPL state manager, and W. B. Edwards, an organizer, spent the night at a hotel in Winlock. The hotel proprietors, joined by a physician and a Methodist minister, later swore affidavits claiming they listened through the wall to the two men discussing the NPL's plan to take over the government, using the farmers and a pretense of democracy as tools. At 2:00 A.M., approximately fifty "businessmen and others" dragged Knutson and Edwards from their room, tarred and feathered Knutson, and warned the men to leave and not stop until they reached Portland. The vigilantes were "said to have been some of the best and most patriotic citizens of Winlock," according to the *Toledo Messenger*.[41]

Knutson apparently left as directed by his attackers, but Edwards was not intimidated. He remained in the area and began calling on farmers near Toledo the next Monday, quickly signing up several. Toledo citizens heard of Edwards' continued presence. Not to be outdone by the people of Winlock and seeing their duty clear, about forty men drove ten miles into the countryside in the dead of night to find him and surrounded the farmhouse where he was sleeping. When Edwards' host, Jens Due, refused to surrender him, the mob threatened to take Due and his son instead. Edwards gave himself up rather than endanger the Due family. He was taken on a lengthy drive, then given his own coat of tar and feathers. "A large party" of Winlock residents drove over to witness the excitement and got their turn with him when the men from Toledo had finished. Edwards now promised to return the money he had collected from local farmers and move on to California. Eight carloads of people escorted him out of the area. The *Toledo Messenger* referred to the vigilantes as "Klu Kluxers," although it is not clear whether they had any formal ties to the Ku Klux Klan, which enjoyed a popular resurgence in the postwar years.[42]

Lewis county Grangers were incensed by the "dastardly" actions of townspeople against peaceful farm organizers, the inaction of law enforcement officials, and the county press's apparent endorsement of vigilantism. The regular quarterly meeting of the County Pomona Grange was scheduled for the week following the incidents. Over three hundred people attended, their interest aroused both by the tar and featherings and by state master William Bouck's appearance as a featured speaker. The meeting began with strongly worded resolutions of patriotism and support for the war effort. Then the Pomona gave a unanimous standing vote to a resolution condemning the mob violence in Winlock and Toledo, attributing it to greed and fear of democracy on the part of war profiteers and calling on elected officials to protect peaceful and lawful assembly. The resolution read:

> Whereas, in the town of Winlock our county farm organizers have been mistreated and brutally beaten by a mob said to have been led by some of the leading citizens of that place and Toledo, and whereas, if these said parties had any knowledge of seditious utterances by these organizers, it was their duty to at once notify the proper government and state officials; but we strongly suspect that the reason the proper officials were not notified, was because they had no proof of such seditious utterances, and farther back of this was the profiteer, making vast profits off the farmers' business, and willing to go any length to prevent the farmers from organizing and helping our president to bring about democracy in the world; therefore be it resolved, that we, the Lewis County Pomona Grange Number 3, in regular session assembled this 3rd day of May, 1918, that the farmers and their leaders are patriotic and doing their best to aid our country in every way to win the war, and that the farmers and their leaders are doing everything in their power to aid the president and his advisors to win the war, and we call upon the businessmen, and good citizens of our county and state to discourage and repudiate the dastardly attempt recently at Winlock and Toledo, to intimidate the farmers of the state by mob violence, and we call upon our county officials and . . . governor of the state, to protect peaceful and patriotic assemblage by the farmers of this state, and our representatives.[43]

This resolution was followed by another which laid much of the blame for mob violence on newspapers throughout Washington and Oregon, for not only failing to represent the interests of farmers but also publishing false accounts and rousing people to a dangerous level of excitement. A new paper was needed in Lewis county, the Grange declared, "which stands for truth, law enforcement and a comprehension of problems of the farmer."[44] Either in an attempt to refute the farmers' stand with an obvious display of impartiality, or out of careless editing, the *Bee Nugget* printed both resolutions in their entirety.

The *Bee Nugget* was well aware of the strength of the Grange locally and unwilling to lose its position as the major county paper by alienating its rural subscribers. Over the next several weeks, its editor tried playing both sides of the fence. It continued its attack on the NPL, reporting an alleged plot to unite the NPL and IWW at the national level and quoting the county prosecutor, who described it as "suspicious that so many of their [NPL] members are suspected of being pro-German." At the same time, the paper printed long repudiations of the NPL by national Grange leaders. The Grange, according to this account, worked for both farm interests and the common good, not selfish class interests, and it was absolutely loyal to the government— in purported contrast to the NPL.[45]

Despite the rough treatment of its organizers, the NPL viewed Lewis county as fertile enough territory to send two more men back in June. They were arrested in Centralia by the county sheriff and charged with inciting to riot. The county prosecutor made clear his belief in the disloyalty of the NPL, but also stated his unwillingness to tolerate more tar and feather parties. The same prosecutor, one year later, was to participate in the planning committee delegated to "cleanse" Centralia of the IWW. Fifteen Lewis county farmers signed bonds ranging from $300 to $5,000 to raise the $23,000 necessary to free the two NPL organizers from jail.[46]

Local events were now overtaken by even more dramatic occurrences elsewhere. William Bouck, as state Overseer, had automatically assumed the mastership of the State Grange at Kegley's death in 1917. Bouck was as dedicated to political reform as Kegley had been, and was himself a member of the Nonpartisan League. At the state convention held in June 1918 in Walla Walla, Bouck stood for election in his own right. Like the Lewis County Pomona, the State Grange began its meeting with rousing resolutions of patriotism, endorsement of the President's actions, and support for the war effort. Officer elections then followed, with Bouck winning fifty-nine percent of the vote. Many outside the organization viewed Bouck's election as an endorsement by the State Grange of the NPL. Walla Walla citizens were primed to respond. At ten o'clock on the evening of June 6, 1918, the State Grange's installation ceremony for their new officers was interrupted by a delegation from a Walla Walla citizens' meeting backed by a number of "husky" gunmen gathered outside. The delegation gave the Grange thirty minutes to vacate the hall. Fifteen minutes of arguments produced no change of heart, so the Grangers closed their meeting in due form,

then marched from the hall carrying the flag and singing "America."[47]
The Grange was unable to find another meeting site in Walla Walla.
The town mayor and sheriff refused to guarantee the safety of the
Grangers. The executive committee sent a telegram to governor Lister,
the former Populist, requesting use of the National Guard armory located there. They received no reply. Committees conducted their business in hotel rooms, and the executive committee finished up the
necessary work in the more hospitable—or at least anonymous—environment of Seattle.[48]

The troubles of Bouck and the state Grange were only beginning. In
August, Bouck was indicted by a Seattle grand jury for violating the
Espionage Act. Although he had expressed his loyalty to the President
and the war effort, Bouck had also publicly disagreed with the policy
of financing the war through the sale of Liberty bonds. He would have
preferred a steep tax on the wealthy. The charges were dismissed before
year's end, but that hardly vindicated him in the eyes of those who saw
him as dangerous. Members of the American Legion visited the 1920
State Grange meeting in the logging town of Aberdeen to assure themselves of the order's loyalty, after "a very misleading article" on the
master's address appeared in the local paper.[49]

Bouck was also attacked by the National Grange, which had long
been opposed to Washington Grange practices, particularly its ties with
organized labor. While the National Grange favored the open shop and
viewed labor and farmer interests as irreconcilably opposed, Bouck and
Kegley before him saw farmers and laborers alike oppressed by industrial capitalists and united as producers. Bouck's views won him frequent praise from Seattle's labor newspaper, the *Union Record.* At the
1920 national meeting in Boston, Bouck was brought up on charges
made by a group of dissident Yakima Grangers that he had injected
partisan politics into the Grange, recruited timber workers and radicals,
and fostered disloyalty to the national body. He was forced to apologize
to the national assembly and pledge on the Bible to uphold the order.
Bouck did not change his ways, however. His master's address at the
1921 state meeting was more radical and inflammatory than ever. Even
some of his long-time supporters thought he had gone too far. The
national executive committee suspended Bouck for failing to live up to
his pledge and ordered him to stand trial at the next national meeting
in Portland, Oregon. At that meeting he was permanently ousted from
the Grange.[50]

Lewis county Grangers steadfastly supported Bouck through the time of his ouster and remained committed to activist politics. Alpha and St. Urban Granges refused to endorse resolutions they received from a subordinate Grange in another county during the summer of 1918 condemning Bouck as unfit for state master and disavowing the Nonpartisan League. Alpha later took an advisory vote for state officers in which Bouck was the overwhelming favorite. After Bouck's arrest that fall, Silver Creek Grange assessed each member one dollar for a defense fund, and contributed another ten dollars for his fight with the National Grange in 1920. Following Bouck's suspension from the order in 1921, the Cowlitz Prairie Grange debated at two different meetings whether to withhold their dues to the State Grange until after the national meeting that was to determine Bouck's status in the organization, but finally decided to send the money in.[51] The Lewis County Pomona invited Bouck to be the featured speaker at their summer picnic and passed a resolution expressing confidence in him and asking the national master for his reinstatement. The Cougar Flat Grange, located just north of Vader, passed its own resolution which it forwarded to the county press for publication, expressing resentment at the National Grange's action and stating: "We will do all in our power to support him [Bouck] in his just fight for reinstatement and the stand he has taken for the redemption of the workers and farmers."[52]

The editors of the Bee Nugget did not share local Grangers' enthusiasm for Bouck. The paper called the reelection of Bouck as state master in June 1921 "past understanding," and expressed satisfaction at his suspension and final expulsion. The editors could not believe that "decent" people could support a man so disgraced and with proven radical associates, and predicted the Grange's downfall. An editorial following the summer Pomona meeting intoned:

Grange picnics and gatherings in various parts of the county make capital of the fact that Bouck, the disgraced head of the organization in this state, and [North Dakota] Governor Lynn J. Frazier, reported one of the most radical of the NPL ill fame, will be the main speakers. Is it any wonder that decent people are rapidly leaving the grange organization in Washington? If it continues to stand by men of this kind, the organization in this state will certainly be outlawed by the great national organization.[53]

Far from leaving the Grange, however, farmers had been joining up in large numbers under Bouck's tenure. Grange membership increased by forty-seven percent in Lewis county between 1918 and 1921, and by thirty-three percent statewide.[54] Lewis county Grangers supported Bouck because they shared his reform vision and rejected the limited version of patriotism that the county press and town vigilantes espoused. They also strongly resented any attempt by outsiders to limit their choice of leaders or dictate their reform agenda. While local Grange meetings throughout this period continued to be occupied largely with initiation of members, social events, and cooperative endeavors, they maintained a consistent interest in political affairs. As in the past, Granges recommended specific measures geared toward increasing democracy, lowering taxes, improving their communities, and creating a more just society. Among the issues discussed by local Granges in 1919 were abolishing capital punishment, stopping a judicial pay raise, and establishing the League of Nations. Improving local roads, government control of railroads, and protecting the right of direct legislation also continued as important issues.[55]

Lewis county Grangers also began taking their own actions in support of the producers' alliance during this period. Prior to the war, cooperation between farmers and urban workers had been most evident among state leaders in lobbying the legislature and campaigning for initiatives. Now the Pomona and subordinate Granges passed their own resolutions declaring their solidarity with labor. In December 1919 Ethel Grange sent a letter of support to striking miners. The next spring in response to a letter from Seattle longshoremen, Alpha Grange agreed to boycott the Charles H. Lilly company because of unfair labor practices. Meeting in Mossyrock that April, the Pomona passed a sweeping prolabor resolution in response to the same issue. The resolution acknowledged the justness of workers' struggles for better conditions and the economic link between good wages for urban labor and good prices for farm goods. The Grangers pledged not knowingly to buy products from any firm that tried to crush labor organization. They resolved:

> Whereas, the organized workers of the cities are continually battling for better working conditions, while many large employers of labor seem determined to crush all attempts on the part of their employees to organize and deal collectively with their employers in matters concerning wages and working conditions; and

Whereas, we recognize that the workers of the cities constitute the greater proportion of the consumers of the products of the farms, and that the worker as a rule spends in direct proportion to his wages, therefore making the wages and conditions of the city workers a matter of vital concern to all farmers,

Therefore, be it resolved, that we hereby pledge our moral support to all organized workers in their legitimate efforts to better their conditions; and we further pledge ourselves that we will not knowingly buy the products of any firm that persists in attempting to undermine the conditions of the working people, and we urge upon all Grangers to lend their support to this matter.[56]

The timing of these resolutions was significant. During the war, striking had been considered tantamount to treason in many quarters and had been portrayed as such in the county press. These resolutions came following the attacks on Nonpartisan League organizers near Winlock, the Seattle General Strike, and the Centralia Massacre, and in the midst of Bouck's ongoing fight with the government and the National Grange. Lewis county Grangers in 1919 and 1920 were going out of their way to take sides in a highly charged atmosphere. They were declaring solidarity with laborers in Seattle and other faraway cities rather than with the townspeople of Winlock, Toledo, and Chehalis. This stance was also in sharp contrast to that taken by the organizers of the Lewis-Pacific Dairymen's Association, who had made a point of declaring small-town businessmen their natural allies. The Grange members voting on these resolutions were well aware of the events of the previous two years and the implications of their statements.

We have no information on how many farmers in Lewis county joined the Nonpartisan League. Given the solid support for Bouck, a League member, and the constant hammering in the Bee Nugget against the League, it is safe to assume that the organization had a fair following. Throughout 1920, the Bee Nugget was constant in its warnings about the League. The paper declared North Dakota close to ruin under NPL leadership, with disastrous tax policies and state ownership of "practically everything" driving out investment capital. The paper's editors also expressed outrage that League organizers asked recruits for cash dues. The Toledo Messenger was also opposed to the League, declaring it intent on taking over the government, and ending one editorial with this warning to League supporters: "Americans will look out for America and woe unto the chap who in the future does not come up to the standards of loyal citizenship."⁵⁷ The *Bee Nugget* pursued the theme of disloyalty as well, claiming the League was un-American and affiliated with the most radical and dangerous "Reds," and that it sought to destroy the American home. One editorial read:

> Any man who is a member of the N-PL who has watched the thing and who is not biased in his view will admit that its teachings are not conducive to good citizenship at all. He will admit that he has gained no benefit from his membership. He will admit that he is losing his respect for the league because of its importation of the most radical "reds" and rabid socialists to further its doctrines. He will admit that he is an American, and no man can say he is giving his best to his country when he allies himself with a man convicted of disloyalty, and one who through the employ of hired agitators seeks the entire destruction of the most sacred American institution—the American home.⁵⁸

The election of 1920 took place in this highly charged climate, with definitions of patriotism and defense of constitutional liberties clearly at issue. Whether or not to form a third party had been an issue of sharp debate among Washington reformers. The old Joint Legislative Committee had not survived the war years, but a new labor-farmer coalition had been formed in 1919 by the major farm organizations in the state, trade unions, and the Railroad Brotherhood. The new Triple Alliance followed the strategy of the Nonpartisan League in attempting to win control of regular party nominations. The producers' alliance was seriously divided, however, by the issue of prohibition, and it began at an inauspicious time. The year 1919 started with the Seattle General Strike and ended with the Centralia Massacre. Both events were disasters from the point of view of the labor unions involved, but struck terror of class warfare and revolution into even moderate Americans otherwise disposed toward reform. By the summer of 1920, many state leaders were ready to try a new approach. In July, four Washington progressive political groups, including the Triple Alliance and the Nonpartisan League, held simultaneous conventions in Yakima. The four organizations quickly agreed on a platform to further democratize election laws, encourage cooperative marketing, allow considerable autonomy for large cities, ban the importation of liquor, exempt farm improvements and modest homes from taxation, and increase support for public schools. The assembled reformers could not agree, however, on a proposal to create a third party affiliated with the recently created national Farmer-Labor party. In the end, a compromise was agreed to. A state Farmer-Labor party was organized to run candidates in western Washington, but in the east, where the Nonpartisan League was strongest, a nonpartisan strategy of capturing nominations would be followed. Bouck participated in the creation of the new party and ran for Congress on its ticket. Like other third parties before it, the Farmer-Labor party ran better than the Democrats in Washington's general election of 1920, but was unable to oust the Republicans. Bouck received forty percent of the vote for Congressman in his district. Robert Bridges, the gubernatorial candidate, ran a poor second to Republican Louis Hart, but better than the Democratic candidate.[59]

Much of the support for the Farmer-Labor party in Washington in 1920 came from small-scale farmers in western Washington, including those in Lewis county. Although the Republicans, draped in a narrowly defined patriotism, were clear winners in both the state and Lewis county, seven of twenty study precincts gave majorities to the Farmer-Labor gubernatorial candidate, Robert Bridges, and an eighth returned a plurality for him. The presidential candidate on the third-party ticket did less well, receiving a majority or plurality in only four of the study precincts. Salmon Creek, Cowlitz Bend, Prescott, Alpha, Ainslie, Veness, Cinebar, and Salkum endorsed Bridges, a long-time activist in state reform politics, for governor. The first four also supported the Farmer-Labor presidential candidate. All were rural precincts with a history of third-party voting. The precincts with the majority of the Finnish population—Ainslie, Veness, and Prescott—continued leaning toward third parties, but local sources are notably silent about any perceived connection between Finns and radicalism, despite the heated rhetoric and the general anti-immigrant bias of the war years.

All of the towns strongly supported the Republican candidates, including Winlock, Vader (formerly Little Falls), Toledo, Napavine, and Onalaska in Granite precinct. A number of agricultural precincts also favored the Republicans, including Cowlitz, Drews Prairie, Eden, Emery (created from rural Napavine precinct), Stillwater, Ethel, and Ferry. Most of these had leaned toward Republican or Democratic candidates in previous elections. The Grange and the farmers' dairy and poultry cooperatives were active in both Farmer-Labor and Republican rural precincts. One could therefore conclude that participation in these organizations was not a decisive factor in voting patterns. On the other hand, even the Republican rural precincts gave about one-third of the vote to Farmer-Labor candidates, compared to less than twenty percent in the towns, and we have no way of knowing the affiliations of individual voters.[60]

Patriotism was clearly an issue in the 1920 campaign. The Chehalis city commissioners expressed their displeasure with the "seditious talk" of a Farmer-Labor candidate campaigning in that town. In the two logging towns of Doty and Onalaska, third-party candidates had speeches interrupted when locomotives with whistles tied down were run up on nearby tracks. The *Bee Nugget* explained these actions by stating that both Doty and Onalaska had been active in war work and had strong American Legion lodges, and thus would not tolerate seditious talk. Nonetheless, the campaign was fairly quiet. The county Democrats had been unable to field a slate of local candidates, so in most county races only Republican and Farmer-Labor nominees ran. Countywide, the Republican candidates won by nearly two-to-one margins. Despite the strength of the party of protest in many rural areas, the population dominance and general conservatism of the towns left little doubt as to the outcome of elections, either locally or statewide. Fervor was therefore unnecessary.[61]

By 1922, the Farmer-Labor party was greatly weakened. Seattle labor leaders were bitterly divided after the fizzling of the General Strike. The Grange, too, was now in disarray. After his ouster from the Grange, Bouck formed his own organization, the Western Progressive Farmers (WPF). State Grange membership fell by forty-one percent between 1921 and 1923, with some joining the WPF and others simply leaving in anger and disenchantment. The WPF was the only farm organization to endorse the Farmer-Labor party for the 1922 elections. The Grange and the Nonpartisan League, as well as the other stalwarts of reform in the state, the State Federation of Labor and the Railroad Brotherhoods, refused to support it. Instead, they joined with the WCTU, the League of Women Voters, and other moderate reform groups in a successful effort to back the Democratic candidate for Senate, Clarence C. Dill, a young Spokane attorney who pledged to assist farmers and work to repeal repressive wartime legislation. Dill's Republican opponent was former Progressive, now turned staunch conservative, Miles Poindexter, who was seeking reelection and had been targeted for defeat by all reform forces in the state.[62]

Although most historians of Washington politics agree that by 1922 the Farmer-Labor party had ceased to be viable, it still retained a fair amount of support among Lewis county farmers.[63] In the study area, three precincts returned majorities for the Farmer-Labor candidate for Senate, and three additional precincts gave the third party pluralities. In over half of the study precincts, the third party drew more than one-quarter of the vote, despite the Grange's backing of the Democratic candidate. Dill was favored in six precincts, and the remaining eight precincts favored the conservative Poindexter, who probably retained some support based on his previously progressive reputation.[64]

In the race for state representative, long-time Alpha Grange lecturer and future president of a three-county Western Progressive Farmers district, Emma Uden, ran on the Farmer-Labor ticket. The Republican candidates won easy victories, however. Uden and the other Farmer-Labor candidates for state representative won big in Uden's home precinct of Alpha, but gained pluralities in only three other precincts, Cowlitz Bend, Eden Prairie, and Salmon Creek. These latter three were adjacent precincts on the southeast bank of the Cowlitz River. The Eden Prairie Grange, among the most severely affected by Bouck's dismissal from the order, was one of four subordinate units out of eleven in the study area to close between 1921 and 1923.[65]

The overall support of Republican candidates did not represent a completely conservative turn. A number of initiatives and referenda were also on the 1922 ballot. Two of the most important from the point of view of reformers were Referenda 14 and 15. The previous year the state legislature had enacted laws that required candidates for office to pledge to support their party platform and required voters to register their party affiliation prior to voting in the primaries. These laws were widely regarded as attempts to eliminate the power of the Nonpartisan League and similar reform groups. The laws were opposed by organized labor, the Grange, the League, and the Farmer-Labor party, and were referred to the voters through the efforts of those organizations in 1922. Despite support by party regulars and the generally conservative bent of voters, the new legislation was decisively defeated at the polls. In the study area, even the most staunchly Republican precincts voted against the two laws, and in ten of the twenty precincts over eighty percent of the vote was cast against the measures. Apparently most voters agreed that the laws were unfair attempts to limit the political process, whatever their feelings toward the Nonpartisan League.[66]

The final strong third-party showing came in 1924. That year Robert LaFollette, a prominent reformer from Wisconsin, ran for President as head of a newly created Progressive party backed by many reformers across the country. In Washington, both Bouck and the Grange's new

state master, Alfred Goss, endorsed LaFollette. Bouck himself had originally won the vice-presidential nomination on the national Farmer-
Labor ticket, but that ticket had been dumped by party insiders in favor
of one headed by Communist leader William Z. Foster. Although Bouck
had remained one of the few noncommunists in the state willing
to work with communists in the increasingly hostile atmosphere of
the early 1920s, he did not back the new Farmer-Labor ticket. A state
Farmer-Labor organization did remain, however, which fielded the local reform slate. LaFollette won in only three of the study precincts,
but he earned over one-third of the vote in eleven of twenty precincts,
and over twenty percent of the vote in all but two. The Farmer-Labor
candidate for governor had a much poorer showing, with no precinct
giving him over one-third of the vote. The Republican candidate won
in all but Alpha precinct.[67]

Unfortunately, little record of the Western Progressive Farmers remains in Lewis county. The Granges that survived Bouck's ouster never
mentioned the new organization in their minutes. The WPF itself was
dormant by the end of the 1920s, and the local groups left no permanent record. The best information we have comes from the official
newspaper of the order, the *Western Progressive Farmer*, published in Seattle beginning in January 1923.

Bouck began setting up his new organization even before his final
ouster by the National Grange. He patterned its structure after the
Grange, even continuing to use the name until forced by lawsuit to
drop it. The WPF was committed to the principle "that all community-
made values belong to the community." It combined endorsement of
a number of reforms that had long been sought by agrarian activists
with the spiritual fervor of Protestant revivalism and a touch of Marx-
ism. The WPF passed resolutions in 1923 supporting public ownership
of the means of transportation and communication, as well as electric
power and water facilities; complete government control of money and
expansion of the money supply; and increases in direct democracy.[68]
The WPF also continued the State Grange's long support of and cooperation with labor. Economic justice was a major concern—that
farmers receive a return on their labor that recognized the basic necessity of the food they produced; that workers receive a wage that
reflected the value of what they produced; and that no one grow rich
off the labor of others. One article explaining the purpose of the WPF
stated, "We will remove from our backs the great load of parasites that
are now enjoying all the luxuries at our expense."[69] In Bouck's eyes,

poverty and injustice would only cease when capitalist exploitation ended. He wrote, "Never will poverty be eliminated till the total wage of the world equals the total market value of the products of industry. Never, till that time, will wars cease, unemployment cease, and democratic government be a reality."[70]

As with many leaders of protest before and since, Bouck believed himself and his organization to be far truer to American ideals than those organizations that claimed patriotism as their main purpose. Far from advocating a "Bolshevik revolution" as many of its detractors claimed, the WPF sought to restore adherence to the Constitution. The continued persecution of the IWW represented a direct assault on the Constitution, according to the WPF paper. Evangeline Douglas of Skagit county wrote condemning the common practice of allowing groups such as the American Legion, the Kiwanas, and Commercial Clubs to speak against radicalism in public high schools. The other side also had a right to be heard, she maintained. Another WPF local passed a resolution condemning the Ku Klux Klan and linking it to Chambers of Commerce. The resolution described the Klan as an "un-American organization imported from Europe and strongly financed by an invisible power for the purpose of causing hatred and trying to divide the working classes."[71] An article surveying attacks on civil liberties throughout American history concluded with this warning for constant vigilance in protecting constitutional rights: "The lesson of this hasty survey is that liberty was never perfect or self-enforcing. It must be fought for daily as we fight for daily bread. We cannot enjoy liberty because the mere words of broad guarantees are in the Constitution. The guarantees are always 'interpreted,' and interpreted by the party in power. We have never enjoyed anything like the absolute liberty which may be deduced from the actual words of the guarantees."[72]

Education of the entire family and the participation of both men and women were also important to the new organization. Bouck's wife, Lura Bouck, headed the juvenile branch of the order, as well as serving as state chaplain, and education was constantly stressed in the paper. Despite the insistence by Bouck that women were as important as men in the organization, the list of state committees for 1923 shows only about one third including women. Women were well represented generally on committees dealing with organizational and political issues. For instance, women chaired committees on taxation, political duties of members, constitution and bylaws, and education. They were absent, however, on those concerned with such practical issues as stock raising, roads, and forestry and conservation.[73] Officers of local units listed in the WPF paper were more often male than female, but women frequently held important posts. Emma Uden, formerly of Alpha Grange, was elected president of a three-county district unit including Lewis, Thurston, and Grays Harbor in May 1923. Her leadership abilities had never been given such recognition in the Grange. Women also served as vice president, lecturer, and secretary in WPF locals.[74] The relative level of equality that women had attained in the State Grange clearly carried over into the WPF, and perhaps the smallness and newness of the organization permitted even more opportunity on occasion. Equally clearly, however, cultural barriers and a sexual division of labor persisted. Other than occasional urgings for participation by women and youth, issues of women's rights were not discussed in the WPF paper. Perhaps with the recent passage of a national women's suffrage amendment, the matter no longer seemed urgent. The State Grange, on the other hand, continued its long pursuit of equality, endorsing the "Lucretia Mott amendment"—the Equal Rights Amendment—in 1924.[75]

The Grange itself came under frequent attack in the WPF press. The national body was a particular target—as it had long been for Washington Grangers. Elihu Bowles, WPF editor, wrote a series of articles entitled "Why I Cannot Support the Grange." His reasons included the national order's support of property rights over the rights of the people; its slow acceptance of women's suffrage; its endorsement of the open shop and antistrike laws; its opposition to government control of railroads and overseas shipping; and its support for persecution, deportation, and execution of radicals. Although the Washington Grange remained dedicated to reform, it too came under attack. Exchanges of speakers between Chambers of Commerce and Granges in both Lewis and Snohomish counties were described as "the great game that is now taking place to bottle up any progressiveness that may remain in farm organizations."[76]

The loss of Bouck and many of the more radical members had not turned the State Grange into a reactionary organization, however. Albert S. Goss was elected new state master at the highly charged state meeting in Yakima in 1922 from a field of six candidates. The sole candidate calling for an end to political involvement was eliminated on the first round of balloting. Goss called for healing the wounds of the order through frankness and a continued emphasis on cooperative economic endeavors, legislative lobbying, education, and social work. He had already gained prominence in the state as manager of the Associ-

ated Grange Warehouse Company, which in the previous three years
had launched a successful network of Grange warehouses and stores
throughout the state.[77] Goss continued the kind of alliance building
around progressive issues that Kegley had so successfully done. In the
state he worked for tax reform and public ownership of power facilities,
and in the National Grange lobbied hard for an increasingly democratic structure and progressive stands on issues. While shunning partisan politics, Goss insisted that the Grange stand by its principles and
continue to be active in legislative work.[78]

The extent of WPF organization in Lewis county is difficult to assess. Four Granges folded in the study area between 1921 and 1923,
although there is no evidence that they converted wholesale to the
new organization. When the three-county WPF district was organized
in May 1923, it included representatives from eight locals. Among
them were Toledo, Alpha, Chehalis, and Centralia.[79] It is possible that
the Toledo and Alpha locals were made up largely from members of
the former Eden Prairie and Cinebar Granges. The only individual
names known from those two locals, however, are Emma and Harry
Uden, both formerly of the still surviving Alpha Grange.

Grange membership in Lewis county had been steadily increasing
during the 1910s, reaching a high of 1,678 members in twenty-three
locals in 1921. Then it dropped by nearly twenty-five percent following
Bouck's ouster, and by an additional twenty-five percent the following
year. Individual Granges were affected differently. Hope officially lost
no members, though attendance was so low during 1922 that they considered disbanding. Alpha declined from thirty-nine to thirty-one members. Included in that loss, no doubt, were the Udens and their three
sons. By 1924, however, membership in Alpha Grange had nearly tri-
pled to ninety-one. Silver Creek and Cowlitz Prairie, on the other hand,
each lost nearly half their members, and remained at the lower levels
through 1926. St. Urban, the largest Grange in the area, declined by
about one-quarter and regained about half that number. Cougar Flat,
which had taken one of the strongest pro-Bouck stances, lost only one-
sixth its members in the year following his dismissal but continued
declining through 1925. In 1924 a new Grange was organized in Cowlitz
Bend, perhaps attracting some of the former Eden Prairie members,
but it remained small, with only thirteen members in 1926.[80]

Despite the withdrawal of many Grange members in Lewis county
and the creation of a WPF structure, the evidence suggests that few
former Grangers joined the new organization. Reports of Lewis county

locals were few in the WPF paper, and the county press never mentioned the organization. The WPF held six Chautauquas in the state in 1924, but none in Lewis county. These Chautauquas were intended as recruitment events, combining radical speakers on political issues and evangelical singing. Lura Bouck wrote new words to old camp songs like "Stand up! Stand up for Jesus." One of her titles was "The Bolshevik Farmers They Call Us," sung to the tune of "My Bonnie Lies Over the Ocean."[81] The fact that none of these events was held in Lewis county suggests that the WPF had no major following there.

The remaining Grangers did continue discussing and debating political issues and specific pieces of legislation. The Cougar Flat master who had been a strong supporter of Bouck continued in office for 1922. He led discussions of the referenda on the state ballot that fall and suggested the Grange begin studying up on government debt and financing. In 1924, Alpha Grange debated, and then took advisory votes on six referendum measures concerning public power, tax policy, and nonpartisan elections that had been sent by the state body. They endorsed the stand taken by the State Grange on all the issues (26 to 4 was the largest margin, 16 to 9 the narrowest) and sent money to support the campaign. The next month Cowlitz Prairie Grange entertained a visitor from Silver Creek who spoke on the same issues, and then appointed a committee to draft a resolution in support of the initiatives.[82]

Despite the deep divisions that tore through south central Lewis county in the postwar years, many people sought to bind together the community again. In some parts of the state where WPF locals were more successful, there were reports of bitter fights over rights to Grange halls, and even of schoolyards being divided to prevent fights among the children of Grange and WPF members.[83] No such struggles are reported in Lewis county. Rather, as in the aftermath of the political battles of the 1890s, people quietly went about their business and sought to rebuild ties of neighborly cooperation. Even farmers and townspeople were able to find common ground on certain issues. One was the burgeoning farmers' cooperative movement that greatly eased the effects of the deep agricultural depression of the early 1920s for all residents of the area. Another was the good roads issue.

Decent farm-to-market roads and travel routes between towns had continued as a major concern throughout the first two decades of the twentieth century. Granges organized road-work parties, town businessmen lobbied the county commissioners for improvements, good-roads associations held meetings and rallies.[84] The proposed Pacific highway running from the major population centers of Puget Sound south through the other Pacific states roused particularly strong emotions. Rather than follow the rail line, the route chosen by the state legislature would angle southeast from Chehalis to Toledo, bypassing Napavine, Winlock, and Vader, before continuing south to Cowlitz county. Boosters of Toledo were ecstatic. The newly reorganized Toledo Commercial Club envisioned streams of tourists and began plans for a big celebration in the spring of 1923, although construction of their leg of the highway was not to take place until later that summer. Residents of the bypassed communities, on the other hand, were furious and determined to win approval for their own highway.[85]

The controversy was a major issue in the race for county commissioner in 1922. J. R. Morton of Napavine ran in the primaries for the Republican nomination, pledging to see that a highway was built connecting Napavine, Evaline, Winlock, and Vader. He received the endorsement of both town businessmen and farmers of the affected communities. A crowd estimated at over three hundred gathered in Mutrie's hall in Winlock in early September. The evening began with a business meeting of the Washington Cooperative Egg and Poultry Association, immediately followed by a rally for Morton. Rex Smith, who as master of Cougar Flat Grange had been a strong supporter of Bouck, led the rally. Among the speakers were J. C. Lawrence, the manager of the coop, who explained how politicians had manipulated the Pacific highway away from its rightful placement. Also on the program were the president of the Napavine bank, a prosperous Winlock lawyer and Republican party activist, a Winlock real estate agent, the master of St. Urban Grange, and a number of farmers, some speaking in Finnish. The evening concluded with a late supper "prepared by the ladies." The excitement and unity in that corridor did not produce a victory for Morton, however. He lost the primary election to Chehalis and Centralia candidates.[86]

The next February, a proposal for a trunk line connecting Napavine, Evaline, Winlock, Vader, and Olequa in Cowlitz county renewed the excitement. J. C. Lawrence succeeded in obtaining a hearing to urge construction of the road before a joint meeting of the state legislative road and bridges committee. The Southern Lewis County Road Association and Southern Lewis County Community Club organized a rally in Winlock and a caravan to Olympia, forty miles to the north, on the day of the hearing. The *Winlock News* carried an announcement of the rally on the front page under a large "Extra!" headline. "Every man and woman should arrange their labors so that they can attend this rally," urged the editor. "Whether southern Lewis county has roads depends on the cooperation of everyone to be benefitted."[87] An estimated five hundred people showed up for this rally, consuming one thousand sandwiches, five hundred donuts, and five hundred cookies while listening to speeches. The assembled crowd agreed to merge the two southern Lewis county booster organizations and unanimously elected M. J. Rarey, St. Urban Grange master, as president. Most of the executive committee members were businessmen of the affected towns. Shops in Napavine, Evaline, Winlock, St. Urban, and Vader were closed for the day and every available automobile requisitioned for the caravan to Olympia. Declaring the proposed road necessary for the prosperity of both farmers and businesses, the *Winlock News* enthused, "The awakening of the community was never so spontaneous. The populace of the southern end of the county left here as one in motor cars bound for legislative halls in Olympia in an almost steady procession." Despite the impressive showing of citizen unity, however, the state legislature did not include the southern Lewis county trunk road in the state budget. The only consolation was that they also left out paving of the Pacific highway through Toledo.[88]

A proposed fruit cannery for Winlock also received enthusiastic support from both area farmers and businessmen in early 1923. It was the effort to secure the cannery that first got farmers and townsmen meeting together and led to the creation of the Southern Lewis County Community Club. Although the cannery fell through, the club continued, hoping to lure new settlers onto still vacant logged-off lands of the region. The *Winlock News* strongly boosted the new organization, and especially the renewal of town and country cooperation it evidenced, observing with some understatement, "A few years ago Winlock had a fine organization operating under the name of the Bungalow Club. When it 'blew up' the old spirit died and the country people and the people in town have drifted farther apart. There has been that lack of co-operation and pull-together spirit so necessary to do the things we need so badly."[89]

It was just such cooperative endeavors between Grangers and business leaders that the WPF condemned. In the eyes of the more radical agrarians, farmers could not maintain their independence and critical stance while collaborating with the agents of their oppressors. Certainly the days of radicalism in the Grange were over. Neither Cougar Flat and St. Urban Grange, those most involved in this particular alliance, nor the other Granges of Lewis county gave up their independence, however. Lewis county farmers steadfastly held to their Granges and refused to join the Farm Bureau in any large numbers. They were willing to join with small town businessmen on specific projects of mutual interest, but not to give up their separate organization and basis of strength. In a rapidly urbanizing state and nation, their influence was small enough already. Their only hope for political victories now lay in their lobbying power. Clearly their votes could be and were routinely swamped by the overwhelming population majorities of the towns and cities. Even with allies among urban laborers, they could not carry an election. Nonetheless, independent farmers' organizations in the form of the Farmers' Alliance and the Grange had been the main source of social ties, political education, economic cooperation, and consciousness-building for three decades. Political repression and vigilante terrorism along with the rapid changes in the national and local economy and society forced changes in rhetoric and programs but did not end the desire for independence, democracy, and prosperity through labor. By 1923, farmers around Winlock were willing to work side by side with the very men reputed to have tarred and feathered farm organizers five years earlier. But they were unwilling to turn over leadership or abandon their organizational base.

৫ৈ 8 ৯৩

Conclusion

Lewis county farmers participated in a movement that sought to reconcile the emerging industrial order with notions of social justice and democratic participation. Admittedly Lewis county residents were marginal participants in the great drama of societal transformation that took place in turn-of-the-century America. Yet their story brings into focus issues that blur beyond easy recognition in studies of leaders and national events. Their movement met with only limited success in conventional terms, but the issues they struggled with as the twentieth century dawned remain central for us today as we face the twenty-first. How do we reconcile economic growth and social justice? How can average citizens have some control of their destinies in a society dominated by big, anonymous corporations? How can the political process be made truly democratic and conflicting interests reconciled in this sprawling, diverse nation?

The agrarian crusade and the rural culture that fed it were embedded in communities that were far more complex than those mythical small towns of our nostalgic fantasies. Farmers and town business owners confronted each other heatedly, sometimes violently, on political issues for over thirty years, but also shared much in common. The president of the county Farmers' Alliance was a charter member of a Masons lodge, alongside the Toledo townsmen reputed to have vandalized a local Alliance hall. In the twentieth century, fraternal lodges, churches, farmers' cooperatives, the Grange, and business associations continued to mediate the tensions of class differences and political conflict, and build common ground through social relationships and issues of regional development. While the farmers of Lewis county remained largely preoccupied with local concerns, the Alliance and the Grange proved to be remarkable vehicles to democratic involvement that yielded creative and transforming results. Average citizens sharing their concerns and debating popular issues were able to transcend narrowly defined economic interests and unite with a broad-based reform coalition to strive toward social justice. Men and women entered the organizations with clear expectations for appropriate gender roles and reproduced the prevailing sexual division of labor, yet they broke with tradition to accord full citizenship status to women and define political activism as a responsibility shared equally by men and women.

Viewed at the grass-roots level of this western county, Populists, Progressives, rural Socialists, and Farmer-Laborites arose from the same community base. Despite the significant economic change that occurred in the region between 1890 and 1920, and despite the ideological debates both nationally and locally among participants in these movements, farmers remained committed to the same basic principles and the same forms of organization. The Farmers' Alliance and Populist party empowered farmers with knowledge of the political economy and with a unity that made them dare to hope that democracy, independence, and prosperity were realizable goals. That knowledge and hope did not disintegrate in 1896 when the national People's party collapsed, nor did it depend on the election of any particular candidate or passage of specific legislation. Rather, at the heart of their vision of democracy and independence was an open and mobilized community that they continued to build in the reenergized Grange.

Lewis county farmers were flexible and practical in pursuing specific objectives, willing to shift their votes among parties and to try different routes to advancing prosperity for themselves and democracy in the broader society. At the same time, they were highly selective. The decisions Lewis county farmers made in flocking to the Grange and economic cooperatives, but rejecting the Farm Bureau and certain initiatives of the Extension Service, were informed political choices. Farmers adopted those aspects of scientific agriculture whose financing and implementation they could control and joined cooperatives managed by their own representatives. They accepted bonuses and other assistance from businessmen's associations in starting their cooperatives, but refused to cede control to outside "experts" or to organizations that seemed geared toward urban and business interests. These decisions were made with the same commitments to education, self-determination, and economic justice that informed their political choices.

In the upper Midwest, other farmers also made the transition from Populism to twentieth-century manifestations of the movement. The Nonpartisan League of North Dakota and the Farmer-Labor party of Minnesota provided similar forums for citizen mobilization and economic empowerment, and at the national level served as political allies with Washington farmers.[1] Throughout much of the South and Southwest, on the other hand, where tenancy, indebtedness, and a racial caste system were basic parts of the economic structure, Populism was effectively buried. In Michael Schwartz's analysis, the strength of the Southern Farmers' Alliance had come from its union of two distinct strands: the old planter class protesting the economic dominance of northern industrialists, and tenant farmers rebelling against the extreme dependency of the crop-lien and share-cropping systems. The internal cleavages here were too strong for the coalition to have much chance of survival. White landowners of the South broke with the Populists by the end of the 1890s. They united instead with merchants to eliminate local democracy, barring black and poor white farmers from the political system and enacting state legislation that would make labor easily exploitable, with the acquiescence of the national parties, the Supreme Court, and the majority of the northern white populace.[2]

Democratic Populism persisted for three decades in Lewis county in part due to the relatively flat class structure of the area. There were certainly disparities in wealth and interests among farmers and between farmers and townspeople, but local structures of exploitation rarely extended to area farmers. No permanent class of landless agricultural labor lived in the community, and few farmers were indebted to local merchants or bankers. Because of their economic independence, numbers, and organizational unity, farmers retained an important political voice that was heard in Olympia. Agrarian radicals could not be decisively defeated in local struggles the way the poorer members of the Alliance had been in the South—either at the county or the state level. Moreover, local townspeople had their own battles to fight for independence from eastern economic domination. Town businessmen were just as determined as farmers that the Northern Pacific Railroad should pay taxes, and perhaps even more concerned about the encroachments of Weyerhaeuser onto county timber lands. Thus, local Republican and Democratic party organizations also at times followed the lead of area farmers rather than their own national platforms.[3]

For most of this period, both farmers and townspeople recognized that they shared a common commitment to their region's prosperity and understood their political differences as legitimate expressions of American democratic values. What changed most decisively in Lewis county was that some influential townspeople broke with that tradition of democratic tolerance after 1917. During the 1890s, the Farmers' Alliance organized Fourth of July celebrations not only in the countryside, but in Toledo and even Chehalis. Townspeople attended despite known and deeply felt political differences, recognizing radical farmers as fellow patriots. Twenty-five years later, farmers in the Grange still proudly waved the American flag and sang patriotic anthems; but when citizens of Winlock and Toledo tarred and feathered Nonpartisan League organizers, and when rifle-toting citizens of Walla Walla broke up a State Grange meeting, the message was clear: the lines of acceptable political behavior had been redrawn to exclude the old Populist forms of participation.

By this time, the more prosperous farmers on the fertile prairies close to towns and rail lines had already begun to drift away from the "producers' alliance" and into political affiliation with townspeople, just as farmers had in other "core" areas in earlier decades. The debate over the meaning of national loyalty and the intense conflict that accompanied it, as well as the interest farmers and townspeople shared in promoting local economic growth, helped speed that transition. It was those on the more marginal lands and in more remote areas who were most likely to stick to the third-party call for a government active in their own interests in the 1920s. The growing populations of towns within Lewis county and cities within the state furthered the decline of third-party strength and gradually closed off the electoral path to reform. Lobbying and pressure-group tactics targeted at the major parties, however, remained viable options and became the dominant form of political protest and influence by the late 1920s.

Economic independence was an ongoing concern of Lewis county farmers, but their political activism did not depend on economic crisis. It emerged, rather, from a long-standing political culture that lived through eras of relative prosperity as well as depression. In the early twentieth century in Lewis county and across much of rural

America, farm prices were relatively strong. The county press, town businessmen's associations, the federal government and Extension Service, and even the leaders of the large farmers' cooperatives all promoted scientific farming, increased production, and the ceding of leadership to businessmen and "experts." Farmers throughout the country were skeptical of the claims of the Extension Service. If Washington farmers were more successful than most in maintaining their independence, it was due in large part to the State Grange, which articulated issues, brought together a strong reform coalition, and provided the community-based forums to develop and sustain an alternative vision of increased democracy and government support of worker and farmer autonomy. The leadership of the State Grange can be given considerable credit for maintaining agrarian activism through the relative prosperity of the Progressive years and in the face of strong counterpressures.

This study has also made clear, however, that the strength of the individual subordinate Granges did not result from the abilities and strategies of state leadership alone but depended on the locals' maintaining a diverse program and using the talents of both men and women. The rallying of Lewis county and Washington state farmers around the Grange and William Bouck in the highly charged atmosphere of the late 1910s was made possible by the deep sense of commitment, loyalty, and trust that had grown from years of shared meals, celebrations, and sick-bed vigils. The community born of the distinctive and complementary contributions of women and men enabled farmers to stand up to mobs who sought to deny them a legitimate place in the American political system.

The Populist spirit may have died more quickly in some parts of the country in part because women were not fully welcomed, thus limiting the development of community solidarity. Unfortunately, however, the ignoring of women and issues of gender by most scholars leaves us with limited evidence. In their studies of Populism, both Schwartz and McNall present evidence that points to the fundamental importance of community but largely ignore the implications. Schwartz quotes one Alliance member in the South who said, "we soon grew tired of lambasting the merchant." That lodge then began holding political and business meetings—which were apparently all male—only once a month and having literary events including women and children the other weeks.[4] Schwartz's only interest in the quotation, however, is in its reference to political education. He does not explore the added social dimension or the potentially weakening effects of separating political and social activities.

McNall, too, does not follow up fully on the implications of his own evidence. He finds that involvement in one Kansas suballiance for which he has complete membership records depended more on the social connections of neighbors and common church membership than on relative economic standing in the community.[5] Yet he seems to consider community-building functions irrelevant once the people joined the Alliance. He is chiefly concerned with class formation and the ability of farmers to perceive their class position. By identifying votes as the only means for farmers to affect the system and thus defining voters as the only people of importance in the movement, he blinds himself to a potentially vital aspect of organizational strength. If it is true, as McNall contends, that women were marginal members of the Alliance in Kansas, and that in some cases men harassed women from the organization, then this could be a major missing element in the structural weakness he claims undermined the Populist movement in that state.

In Lewis county, the situation was quite different. There, while women constituted a minority of the membership of the Alliance and the Grange, they were welcomed. They actively participated in maintaining the organizations through gender-specific contributions, furthered political debates, and successfully pushed for full inclusion in political affairs. In a society where a gendered division of labor was a basic part of life, women's activities were vital to maintaining a community organization. Their contributions in large part created and maintained the conditions under which political education and class formation, as McNall defines them, could take place. Thus, if men in Kansas and much of the South excluded women from the Alliance, this is an important part of the local power struggle that warrants serious attention. Poor whites in the South worked against their own interests in clinging to white supremacy and thereby assured the continued dominance of elites. Perhaps in a similar manner farm men's exclusion of women from political participation in the Midwest and South helped cement the dominance of the urban core.

McNall misses an important aspect of his analysis because, like many other scholars, he is blind both to considerations of gender and to the possibility of women as serious political actors outside of traditional male bounds. He conscientiously uses inclusive terminology, always saying "men and women" where appropriate, and quotes at length from important women speakers and theorists, but he ignores women when they were not behaving in traditionally defined "male" political ways. McNall does not examine those rural neighborhoods and churches which recruited members, or the gendered division of labor which was a part of their structure and was incorporated into the Alliance.

Were there differences in the gender structures of Kansas, Georgia, Lewis county, and other sites of agrarian activism that furthered or hindered strong organizational growth? The limited research and the difficulty in finding sources that give a clear picture of household and community organization together make it impossible to say at this point. Surely there were local differences, rooted in different economic structures and cultures. The link between white supremacy and male dominance in the South has been noted frequently.[6] Yet there is also evidence of a broadly similar rural culture existing in at least the northern part of the country around the turn of the century. Studies of farming areas in the Northeast and the Midwest reveal attitudes toward work, family, and gender consistent with each other and with the limited evidence uncovered in this study—and distinct from middle-class urban attitudes and values.[7] This culture recognized that the labor of women in the household economy was important, even if it did not always value women's labor as highly as men's. The culture also gave women a central role in community life. Women's contributions were often gender-specific and sharply limited, but nonetheless were made in a sexually integrated public arena and were vital to the long-term success of community organizations. Moreover, once they were part of an organization that advocated the equality of women even if it did not always practice it, peoples' understanding of the gender structure of their society was open to the same kinds of changes as occurred in other aspects of their political and economic analyses.

The Grange in Lewis county remained an organization truly of and for farmers, and one that also recognized and furthered women's leadership abilities in a mixed-sex setting. In doing so, it continued a practice pioneered in the Farmers' Alliance three decades earlier, and one that remained unique among Lewis county organizations in the 1920s. The structure of the Winlock area's Southern Lewis County Roads Association of the early 1920s was telling. Women clearly attended the rallies and were welcomed as enthusiastic members, but the expected nature of their role was also quite clear. They made sandwiches and cookies and helped swell the crowd. The gender structure followed that of the fraternal lodges, by and large. Women served domestic functions but did not speak or serve as officers or on the executive committee.[8] As voters, they were now included in the good-roads meetings—in contrast to the 1890s, when the women had stayed home and the men had had to do without food at their meetings. Without a larger theory of equality, however, or a separate organizational base to empower them, women were incorporated into the good-roads movement in accordance with prevailing gender norms and without a mechanism to challenge them.

Lewis county farmers had maintained their own organizations for thirty years. As with urban women's associations, the Alliance and the Grange had given farmers a consciousness of their common problems, practice in political participation, and an agenda for change. By practicing democracy within their own communities, they played a major role in changing the nature of politics in their society. Grangers did not make the mistake that Alliance members before them had of merging their own organization with that of the larger reform coalition. It remained rooted in the core of its strength—the rural community—and so was able to endure. At the same time, the integration of the Grange and its members into the local community structure of diverse organizations required its members to exercise a neighborly toleration of differences. Much as people might have desired and fought for political and social change, they remained committed to getting along with their neighbors, and that required above all compromise and forgiveness.

Notes

1. Introduction

1. The strongest proponent of this view is Lawrence Goodwyn, *Democratic Promise: The Populist Moment in America* (New York: Oxford University Press, 1976), xx. See also Michael Schwartz, *Radical Protest and Social Structure: The Southern Farmers' Alliance and Cotton Tenancy, 1880–1890* (New York: Academic Press, 1976), 14–15; Theodore R. Mitchell, *Political Education in the Southern Farmers' Alliance, 1887–1900* (Madison: University of Wisconsin Press, 1987), chap. 2; Donna A. Barnes, *Farmers in Rebellion: The Rise and Fall of the Southern Farmers' Alliance and People's Party in Texas* (Austin: University of Texas, 1984), 107.

2. Alan Dawley, *Struggles for Justice: Social Responsibility and the Liberal State* (Cambridge: Belknap Press of Harvard University Press, 1991), 1–5. For additional discussion of social struggle and state expansion during this period, see Stephen Skowronek, *Building a New American State: Expansion of National Administrative Capacities, 1877–1920* (Cambridge: Cambridge University Press, 1982); Morton Keller, *Affairs of State: Public Life in Late 19th Century America* (Cambridge: Belknap Press of Harvard University Press, 1977); Walter Dean Burnham, *Critical Elections and the Mainsprings of American Politics* (New York: Norton, 1970) 6–31; Margaret Weir, Ann Shola Orloff, and Theda Skocpol, "Understanding American Social Politics," in Weir, Orloff, and Skocpol, eds., *The Politics of Social Policy in the United States* (Princeton: Princeton University Press, 1988), 3–27.

3. The West as colony of the federal government and economically dependent on the industrial Northeast has long been an issue for western historians, and is a widely accepted notion in recent works. See, for example, Donald Worster, *Under Western Skies: Nature and History in the American West* (New York: Oxford University Press, 1992), 225; Patricia Nelson Limerick, *The Legacy of Conquest: The Unbroken Past of the American West* (New York: Norton, 1988), 78–96; John Walton, *Western Times and Water Wars: State, Culture, and Rebellion in California* (Berkeley: University of California Press, 1992), xvii; Richard White, *"It's Your Misfortune and None of My Own": A New History of the American West* (Norman: University of Oklahoma Press, 1991), 57–59, 236–68; Carlos A. Schwantes, *The Pacific Northwest: An Interpretive History* (Lincoln: University of Nebraska Press, 1989), 14–16; Richard Franklin Bensel,

Sectionalism and American Political Development 1880–1980 (Madison: University of Wisconsin Press, 1984), 18–21.

4. Richard M. Valelly, *Radicalism in the States: The Minnesota Farmer-Labor Party and the American Political Economy* (Chicago: University of Chicago Press, 1989), 10–15.

5. C. Vann Woodward, *Origins of the New South 1877–1913* (Baton Rouge: Louisiana State University Press, 1980 [1951]), chap. 9; Schwartz, *Radical Protest*, 7–11; Steven Hahn, *The Roots of Southern Populism: Yeoman Farmers and the Transformation of the Georgia Upcountry, 1850–1890* (New York: Oxford University Press, 1983), 201.

6. John D. Hicks, *The Populist Revolt: A History of the Farmers' Alliance and the People's Party* (Lincoln: University of Nebraska Press, 1961 [1931]), 237.

7. Hicks, *Populist Revolt*, 2, 406–9. Similar views of Populist influence on Progressive politics are expressed in Stanley B. Parsons, Karen Toombs Parsons, Walter Killilae, Beverly Borgers, "The Role of Cooperatives in the Development of the Movement Culture of Populism," *Journal of American History* 69 (March 1983): 866–85; Gene Clanton, *Populism: The Humane Preference in America, 1890–1900* (Boston: Twayne, 1991), xiv; Robert F. Durden, *The Climax of Populism: The Election of 1896* (Lexington: University of Kentucky Press, 1965), vii, 170.

8. Richard Hofstadter, *The Age of Reform: From Bryan to FDR* (New York: Vintage, 1955).

9. For example, see Norman Pollack, *The Populist Response to Industrial America: Midwestern Populist Thought* (Cambridge: Harvard University Press, 1962); Walter T. K. Nugent, *The Tolerant Populists: Kansas Populism and Nativism* (Chicago: University of Chicago Press, 1961); Michael Paul Rogin, *The Intellectuals and McCarthy: The Radical Specter* (Cambridge: MIT Press, 1967); Thomas Wayne Riddle, *The Old Radicalism: John R. Rogers and the Populist Movement in Washington* (New York: Garland, 1991); Robert Donald Saltvig, "The Progressive Movement in Washington" (Ph.D. diss., University of Washington, 1966); John David Dibbern, "Grass Roots Populism: Politics and Social Structure in a Frontier Community" (Ph.D. diss., Stanford University, 1980).

10. Goodwyn, *Democratic Promise*, xi–xxi.

11. Parsons et al., "The Role of Cooperatives"; Robert C. McMath, Jr., "Sandy Lands and Hogs in the Timber: (Agri)cultural Origins of the Farmers' Alliance in Texas," in Steven Hahn and Jonathan Prude, eds., *The Countryside in the Age of Capitalist Transformation: Essays in the Social History of Rural America* (Chapel Hill: University of North Carolina Press, 1985); Scott G. McNall, *The Road to Rebellion: Class Formation and Kansas Populism, 1865–1900* (Chicago: University of Chicago Press, 1988).

12. McMath's most recent work gives an excellent overview of the Alliance and Populist movements, incorporating much recent scholarship in rural social history: Robert C. McMath, Jr., *American Populism: A Social History, 1877–1898* (New York: Hill and Wang, 1993). His earlier work also expresses this view: Robert C. McMath, *Populist Vanguard: A History of the Southern Farmers' Alliance* (Chapel Hill: University of North Carolina Press, 1975), chap. 10; McMath, "Sandy Lands and Hogs," 206.

13. See Schwartz, *Radical Protest*; Barnes, *Farmers in Rebellion*; McNall, *Road to Rebellion*. For a sampling of relevant discussions of resource mobilization theory, see Bert Klandermans and Sidney Tarrow, "Mobilization into Social Movements: Synthesizing European and American Approaches," in *International Social Movement Research*, vol. 1 (Greenwich, Conn.: JAI Press, 1988), 1–38; Charles Tilly, "Social

Movements, Old and New," in *Research in Social Movements, Conflict, and Change*, vol. 10 (Greenwich: JAI Press, 1988), 1–18; Mayer N. Zald, "The Trajectory of Social Movements in America," *Research in Social Movements, Conflicts, and Change*, vol. 10 (Greenwich: JAI Press; 1988); 19–41; Mayer N. Zald and John D. McCarthy, eds., *Social Movements in an Organizational Society: Collected Essays* (New Brunswick, N.J.: Transaction Books, 1987); Dennis Chong, *Collective Action and the Civil Rights Movement* (Chicago: University of Chicago Press, 1991); Howard Kimeldorf, *Reds or Rackets?: The Making of Radical and Conservative Unions on the Waterfront* (Berkeley: University of California Press, 1988).

14. For works on twentieth-century agrarian activism, see James R. Green, *Grass-Roots Socialism: Radical Movements in the Southwest 1895–1943* (Baton Rouge: Louisiana State University Press, 1978); Charles Edward Russell, *The Story of the Non-Partisan League: A Chapter in American Evolution* (New York: Harper, 1920); Millard L. Gieske, *Minnesota Farmer-Laborism: The Third-Party Alternative* (Minneapolis: University of Minnesota Press, 1979); Kathleen Diane Moum, "Harvest of Discontent: The Social Origins of the Nonpartisan League, 1880–1922" (Ph.D. diss. University of California, Irvine, 1986); Garin Burbank, *When Farmers Voted Red: The Gospel of Socialism in the Oklahoma Countryside, 1910–24* (Westport, Conn.: Greenwood, 1976); Theodore Saloutos and John D. Hicks, *Agricultural Discontent in the Middle West, 1900–1939* (Madison: University of Wisconsin Press, 1951).

15. See, for example, the exchange between Lawrence Goodwyn and James Green: Lawrence Goodwyn, "The Cooperative Commonwealth and Other Abstractions: In Search of a Democratic Promise," *Marxist Perspectives* 3 (Summer 1980): 8–42; James Green, "Populism, Socialism, and the Promise of Democracy," *Radical History Review* 24 (Fall 1980): 7–40; Green, *Grass-Roots Socialism*. Rogin, *Intellectuals and McCarthy* is one of the few books to integrate discussions of Populism and twentieth-century movements. See also Theodore Saloutos, *Farmer Movements in the South, 1865–1933* (Lincoln: University of Nebraska Press, 1960).

16. Paula Baker, "The Domestication of Politics: Women and American Political Society, 1780–1920," *American Historical Review* 89 (1984): 620–47; Paula Baker, *The Moral Frameworks of Public Life: Gender, Politics, and the State in Rural New York, 1870–1930* (New York: Oxford University Press, 1991), xvii. For supportive arguments, see also Nancy F. Cott, *The Bonds of Womanhood: "Woman's Sphere" in New England, 1780–1835* (New Haven: Yale University Press, 1977); Ruth Bordin, *Woman and Temperance: The Quest for Power and Liberty, 1873–1900* (Philadelphia: Temple University Press, 1981); Barbara Leslie Epstein, *The Politics of Domesticity: Women, Evangelism, and Temperance in Nineteenth-Century America* (Middletown, Conn.: Wesleyan University Press, 1981); Karen J. Blair, *The Clubwoman as Feminist: True Womanhood Redefined, 1868–1914* (New York: Holmes & Meier, 1980); Linda K. Kerber, "Separate Spheres, Female Worlds, Woman's Place: The Rhetoric of Women's History," *Journal of American History* 15 (June 1988): 9–39; Peggy Pascoe, *Relations of Rescue: The Search for Female Moral Authority in the American West, 1874–1939* (New York: Oxford University Press, 1990).

17. For examples of challenges to the primacy of separate spheres and examples of the diversity among women activists, see Ellen Carol DuBois and Vicki L. Ruiz, "Introduction," in DuBois and Ruiz, eds., *Unequal Sisters: A Multi-Cultural Reader in U.S. Women's History* (New York: Routledge, 1990); Nancy A. Hewitt, *Women's Activism*

and Social Change: Rochester, New York, 1822–1872 (Ithaca: Cornell University Press, 1984); Nancy F. Cott, *The Grounding of Modern Feminism* (New Haven: Yale University Press, 1987); Elsa Barkley Brown, "Womanist Consciousness: Maggie Lena Walker and the Independent Order of St. Luke," *Signs: Journal of Women in Culture and Society* 14 (1989): 610–33; Paula Giddings, *When and Where I Enter: The Impact of Black Women on Race and Sex in America* (New York: Bantam Books, 1985); Maurine Weiner Greenwald, "Working Class Feminism and the Family Wage Ideal: The Seattle Debate on Married Women's Right to Work, 1914–1920," *Journal of American History* 76 (1989): 118–49; Iris Berger, Elsa Barkley Brown, and Nancy A. Hewitt, "Symposium: Intersections and Collision Courses: Women, Blacks, and Workers Confront Gender, Race, and Class," *Feminist Studies* 2 (Summer 1992): 283–326.

18. Nancy Grey Osterud, *Bonds of Community: The Lives of Farm Women in Nineteenth-Century New York* (Ithaca: Cornell University Press, 1991), 1–2, 150–58; Mary Neth, "Preserving the Family Farm: Farm Families and Communities in the Midwest, 1900–1940" (Ph.D. diss., University of Wisconsin, 1987); Karen V. Hansen, " 'Helped Put in a Quilt': Men's Work and Male Intimacy in Nineteenth-Century New England," *Gender and Society* 3 (September 1989): 334–54; Jane Taylor Nelsen, *A Prairie Populist: The Memoirs of Luna Kellie* (Iowa City: University of Iowa Press, 1992). The heavy burdens of farm women's economic roles are highlighted in Deborah Fink, *Agrarian Women: Wives and Mothers in Rural Nebraska, 1880–1940* (Chapel Hill: University of North Carolina Press, 1992); Joan M. Jensen, *With These Hands: Women Working on the Land* (Old Westbury, N.Y.: The Feminist Press, 1981); John Mack Faragher, *Sugar Creek: Life on the Illinois Prairie* (New Haven: Yale University Press, 1986). For discussion of the rural gender division of labor in an earlier time period, see Laurel Thatcher Ulrich, *A Midwife's Tale: The Life of Martha Ballard, Based on Her Diary, 1785–1812* (New York: Vintage, 1990); Christopher Clark, *The Roots of Rural Capitalism: Western Massachusetts, 1780–1860* (Ithaca: Cornell University Press, 1990); Joan M. Jensen, *Loosening the Bonds: Mid-Atlantic Farm Women, 1750–1850* (New Haven: Yale University Press, 1986). For changing sexual divisions of farm labor in the twentieth century see Deborah Fink, *Open Country Iowa: Rural Women, Tradition, and Change* (Albany: SUNY Press, 1986); Carolyn E. Sachs, *The Invisible Farmer: Women in Agricultural Production* (Totowa, N.J.: Rowman & Allanheld, 1983); Katherine Jellison, *Entitled to Power: Farm Women and Technology, 1913–1963* (Chapel Hill: University of North Carolina Press, 1993); Jane Marie Pederson, *Between Memory and Reality: Family and Community in Rural Wisconsin, 1870–1970* (Madison: University of Wisconsin Press, 1992); Sonya Salamon, *Prairie Patrimony: Family, Farming, and Community in the Midwest* (Chapel Hill: University of North Carolina Press, 1992); Janet M. Labrie, "The Depiction of Women's Field Work in Rural Fiction," *Agricultural History* 67 (Spring 1993): 119–33.

19. McNall, *Road to Rebellion*, 238–44.

20. McMath, *Populist Vanguard*, chap. 5; Julie Roy Jeffrey, "Women in the Southern Farmers' Alliance: A Reconsideration of the Role and Status of Women in the Late Nineteenth Century South," *Feminist Studies* 3 (1975): 72–90; Maryjo Wagner, "Farms, Families, and Reform: Women in the Farmers' Alliance and Populist Party" (Ph.D. diss., University of Oregon, 1986); William C. Pratt, "Radicals, Farmers, and Historians: Some Recent Scholarship about Agrarian Radicals in the Upper Midwest," *North Dakota History* 52 (1985): 12–24; Pauline Adams and Emma S. Thorn-

ton, *A Populist Assault: Sarah E. Van de Vort Emery on American Democracy, 1862–95* (Bowling Green, Ohio: Bowling Green State University Popular Press, 1982); Michael L. Goldberg, "Non-Partisan and All-Partisan: Rethinking Woman Suffrage and Party Politics in Gilded Age Kansas," *Western Historical Quarterly* 25 (Spring 1994): 21–44.

21. For some mention of women, see Green, *Grass-Roots Socialism*, 93–94; Karen Starr, "Fighting for a Future: Farm Women of the Non-Partisan League," *Minnesota History* 48 (1983): 255–62.

22. Karen Brodkin Sacks, *Caring by the Hour: Women, Work, and Organizing at Duke Medical Center* (Urbana: University of Illinois Press, 1988), 120–21. The necessity of understanding gender relations in studying labor movements is also argued forcefully by Elizabeth Faue, *Community of Suffering and Struggle: Women, Men, and the Labor Movement in Minneapolis, 1915–1945* (Chapel Hill: University of North Carolina Press, 1991); Robin D. G. Kelley, *Hammer and Hoe: Alabama Communists during the Great Depression* (Chapel Hill, University of North Carolina Press, 1990).

2. Rural Community Life: Lewis County in the 1890s

1. Alma Nix and John Nix, eds., *The History of Lewis County Washington* (Chehalis, Wash.: Lewis County Historical Society, 1985), 186; *The Toledo Community Story, 1853–1953*, 121 (Seattle Public Library); U.S. Census manuscripts for Lewis county, 1900 and 1910.

2. Because voting precincts and census tracts followed the same boundaries throughout Lewis county and roughly coincided with neighborhood boundaries as well, this study relied on those boundaries. Between 1890 and 1920 a number of new precincts were added as population in the area grew. In 1900 the precincts included were Ainslie, Alpha, Cinebar, Cowlitz, Drews Prairie, Eden, Ethel, Ferry, Granite, Little Falls, Napavine, Prescott, Salkum, Salmon Creek, Toledo, Veness, and Winlock. The area was further subdivided over the next several years to include Cowlitz Bend, Emery, and Stillwater.

3. Nix and Nix, eds., *History of Lewis County*, 267. For documentation of similar burning practices in the Puget Sound area, see Richard White, *Land Use, Environment, and Social Change: The Shaping of Island County, Washington* (Seattle: University of Washington Press, 1980), 20–21.

4. Robert H. Ruby and John A. Brown, eds., *A Guide to the Indian Tribes of the Pacific Northwest* (Norman: University of Oklahoma Press, 1986), 69–71.

5. General overviews of this period can be found in Carlos A. Schwantes, *The Pacific Northwest: An Interpretive History* (Lincoln: University of Nebraska Press, 1989), 28–32, 36–37, 59–63; Gordon B. Dodds, *The American Northwest: A History of Oregon and Washington* (Arlington Heights, Ill.: Forum Press, 1986), 3–12, 35–47; Nix and Nix, eds., *History of Lewis County*, 6–8, 20–21. See also Robert H. Ruby and John A. Brown, *The Chinook Indians: Traders of the Lower Columbia* (Norman: University of Oklahoma Press, 1976); Malcolm Clark, Jr., *Eden Seekers: The Settlement of Oregon, 1818–1862* (Boston: Houghton Mifflin, 1981), 25–56.

6. Ruby and Brown, *Guide to the Indian Tribes*, 70–71; *Toledo Community Story*, 8, 74, 76, 103, 108; Schwantes, *The Pacific Northwest*, 117–19. One woman's experience

of pioneer life and white-Indian relations in the 1850s in neighboring Thurston county is provided in Phoebe Goodell Judson, *A Pioneer's Search for an Ideal Home* (Lincoln: University of Nebraska Press, 1984), 85–193.

7. Wallace J. Miller, *Southwestern Washington: Its Topography, Climate, Resources, Productions, Manufacturing Advantages, Wealth, and Growth* (Olympia: Pacific Publishing, 1890), 113; Citizens of Centralia, *Centralia's First Century 1845–1955*, (Tumwater, Wash.: H.J. Quality Printing, 1979), 35; Thirteenth Census, 1910, 3: 577.

8. Nix and Nix, eds. *History of Lewis County*, 8; *People's Advocate*, May 20, 1898.

9. *Chehalis Bee*, September 5, 1890, October 2, 1891; *People's Advocate*, May 20, 1898; Miller, *Southwestern Washington*, 139.

10. *Chehalis Bee*, October 31, 1890; *Chehalis Nugget*, August 25, 1893; *People's Advocate*, February 16, 1894.

11. Nix and Nix, eds., *History of Lewis County*, 8; *Peoples' Advocate*, May 20, 1898; *Chehalis Bee*, October 2, 1891, November 27, 1891, September 9, 1892.

12. *Chehalis Bee*, July 15–29, 1892, September 2, 1892, January 17, 1896, January 15, 1897; C. C. Wall, "A History of Winlock, Washington," 1952, 44, 52, Lewis County Historical Society, Chehalis, Wash.

13. *People's Advocate*, May 20, 1898; *Toledo Community Story*, 124–47; Nix and Nix, eds., *History of Lewis County*, 9–10.

14. *People's Advocate*, February 17, 1893, March 3, 1893; R. L. Polk & Co., *Lewis County Directory, 1904* (Seattle: Acme Publishers, 1904), 99, 106–7.

15. See U.S. Census summary population volumes for minor civil divisions, 1890–1920.

16. For an excellent discussion of the difficulties of farming logged-off lands in western Washington, see Richard White, *Land Use*, chap. 3.

17. These figures and many of those that follow were derived from the manuscripts of the 1900 U.S. Census for the precincts of the study area. For the purposes of this study, I included everyone age sixteen or over. For more detailed data and methodological discussion, see Marilyn P. Watkins, "Political Culture and Gender in Rural Community Life: Agrarian Activism in Lewis County Washington, 1890–1925" (Ph.D. diss., University of Michigan, 1991), app. B.

18. U.S. Census, 1910, *Abstract with Supplement for Washington*, 632.

19. U.S. Census, *Statistics of Agriculture*, 1890 and 1900; *People's Advocate*, December 23, 1898.

20. Such divisions of family labor at the turn of the century are documented in Nancy Grey Osterud, *Bonds of Community: The Lives of Farm Women in Nineteenth-Century New York* (Ithaca: Cornell University Press, 1991); Mary Neth, "Preserving the Family Farm: Farm Families and Communities in the Midwest, 1900–1940" (Ph.D. diss., University of Wisconsin, 1987); Joan M. Jensen, *With These Hands: Women Working the Land* (Old Westbury, N.Y.: The Feminist Press, 1981); J. Sanford Rikoon, *Threshing in the Midwest, 1820–1940: A Study of Traditional Culture and Technological Change* (Bloomington: Indiana University Press, 1988); Jane Taylor Nelsen, ed., *A Prairie Populist: The Memoirs of Luna Kellie* (Iowa City: University of Iowa Press, 1992).

21. *Toledo Community Story*, 58–61, 83; U.S. Census manuscripts for Lewis county, 1900. Census workers in 1900 were given explicit instructions to record no occupation for wives and daughters living at home and performing household duties without pay. Children working on their parents' farm were to be recorded as farm

laborers, but daughters doing "women's" farm work were apparently considered to be performing "household duties" rather than "farm labor." Only women who managed or supervised the home farm were to be recorded as farmers. In 1910, instructions to recorders changed somewhat, indicating that women who performed outdoor work for their husband or son should be listed as farm laborers, although the practices of Lewis county recorders changed little. See Bureau of the Census, *200 Years of U.S. Census Taking: Population and Housing Questions, 1790–1990* (Washington, D.C.: U.S. Government Printing Office, 1989), 43–44, 52.

22. *Toledo Community Story*, 79–80; U.S. Census manuscripts, 1900.

23. They worked in a legal and ideological setting that tended to devalue their contributions, but farm women took pride in their labor during the nineteenth and early twentieth centuries. See Neth, "Preserving the Family Farm," 355, 367–68, 382, 232–36; Osterud, *Bonds of Community*, 139–47; Karen V. Hansen, " 'Helped Put in a Quilt': Men's Work and Male Intimacy in Nineteenth-Century New England," *Gender and Society* 3 (September 1989): 334–54; Deborah Fink, *Open Country, Iowa: Rural Women, Tradition, and Change* (Albany: SUNY Press, 1986); Joan M. Jensen, *Loosening the Bonds: Mid-Atlantic Farm Women, 1750–1850* (New Haven: Yale University Press, 1986); Jensen, *With These Hands*, 205–7; Katherine Jellison, *Entitled to Power: Farm Women and Technology, 1913–1963* (Chapel Hill: University of North Carolina Press, 1993), 18. The grueling nature of work and the gender inequalities are emphasized more strongly in other writings, including Deborah Fink, *Agrarian Women: Wives and Mothers in Rural Nebraska, 1880–1940* (Chapel Hill: University of North Carolina Press, 1992), 2–5; Carolyn E. Sachs, *The Invisible Farmer: Women in Agricultural Production* (Totowa, N.J.: Rowman & Allanheld, 1983); John Mack Faragher, *Sugar Creek: Life on the Illinois Prairie* (New Haven: Yale University Press, 1986), 118; Eliza W. Farnham, *Life in Prairie Land* (New York: Arno Press, 1972 [c. 1846]).

24. Nix and Nix, eds., *History of Lewis County*, 141, 223–24, 356.

25. The towns of Little Falls and Napavine were not yet incorporated at the time of the 1900 census, so precinct boundaries included a substantial amount of open countryside. These percentages exclude farmers so as to roughly approximate town population.

26. *Chehalis Bee*, February 6, 1896, July 3, 1896, September 24, 1897; *Chehalis Nugget*, August 17, 1894; Wall, "History of Winlock," 45.

27. *People's Advocate*, May 20, 1898, January 6, 1899; U.S. Census manuscripts for Napavine, 1900.

28. *People's Advocate*, October 19, 1900. For histories of the lumber industry at the turn of the century, see William G. Robbins, *Hard Times in Paradise: Coos Bay, Oregon, 1850–1986* (Seattle: University of Washington Press, 1988); Norman H. Clark, *Mill Town: A Social History of Everett, Washington, from Its Earliest Beginnings on the Shores of Puget Sound to the Tragic and Infamous Event Known as the Everett Massacre* (Seattle: University of Washington Press, 1970); Robert E. Ficken, "Weyerhaeuser and the Pacific Northwest Timber Industry, 1899–1903" in G. Thomas Edwards and Carlos A. Schwantes, eds., *Experiences in a Promised Land* (Seattle: University of Washington Press, 1986), 139–52; White, *Land Use, Environment, and Social Change*, Andrew Mason Prouty, *More Deadly Than War: Pacific Coast Logging 1827–1981* (New York: Garland, 1985); Jeremy W. Kilar, *Michigan's Lumbertowns: Lumbermen and Laborers in*

Saginaw, Bay City, and Muskegon, 1870–1905 (Detroit: Wayne State University Press, 1990).

29. U.S. Census manuscripts for Napavine, Winlock, and Cowlitz precincts, 1900. Men whose occupations were recorded as mill worker, timber worker, sawyer, timber feller, logger, engineer, mill owner, and the like were included in these statistics.

30. For discussions of the blurriness and fluidity of class in similar communities, see John Walton, *Western Times and Water Wars: State, Culture, and Rebellion in California* (Berkeley: University of California Press, 1992), 61–86; Scott G. McNall, *The Road to Rebellion: Class Formation and Kansas Populism, 1865–1900* (Chicago: University of Chicago Press, 1988), 140; Robert R. Dykstra, *The Cattle Towns* (New York: Knopf, 1968), 109–10.

31. *Chehalis Bee*, February 9, 1894, February 23, 1894, March 9, 1894, December 31, 1897, January 7, 1898, January 28, 1898, February 4, 1898; *People's Advocate*, February 2, 1894.

32. *Chehalis Bee*, March 11, 1892, April 17, 1896, February 4, 1898; *People's Advocate*, July 9, 1897.

33. Osterud, *Bonds of Community*, 2, 139. See also Neth, "Preserving the Family Farm," 44–45; Nelson, *Prairie Populist*. Farm women isolated by the vast distances of the Great Plains had limited access to such socializing. See Fink, *Agrarian Women*; Melody Graulich, "Violence against Women: Power Dynamics in Literature of the Western Family," in Susan Armitage and Elizabeth Jameson, eds., *The Women's West* (Norman: University of Oklahoma Press, 1987).

34. *Chehalis Nugget,* July 10, 1891, July 24, 1891, August 14, 1891, August 28, 1891.

35. *Chehalis Bee*, April 17, 1896.

36. *People's Advocate*, March 17, 1893, January 5, 1894.

37. *Chehalis Bee*, December 6, 1895.

38. Ibid., June 4, 1897, June 11, 1897.

39. Minnie Lingreen and Priscilla Tiller, *Hop Cultivation in Lewis County, Washington, 1888 to 1940: A Study in Land Use Determinants* (Centralia, Wash.: 1981) provides an interesting and detailed account of hop growing in Lewis county.

40. *People's Advocate*, December 23, 1892, August 2, 1895; *Bee Nugget*, August 12, 1910, September 14, 1911.

41. See Lingreen and Tiller, *Hop Cultivation*, 22.

42. *People's Advocate*, September 24, 1897.

43. Wall, "History of Winlock"; *Toledo Community Story*, 109, 129, 134; Nix and Nix, eds., *History of Lewis County*, 48.

44. Spencer S. Sulliger to Alpha J. Kynett, December 19, 1895, Mayfield Methodist Church, Pacific Northwest Methodist Archives, University of Puget Sound, Tacoma, Wash.; Mossyrock Grange #355, *Memories from Family Albums of School District #206* (Centralia, Wash.: Lightning Print, 1976), 56.

45. See the Church Record of the Winlock Methodist Episcopal Church, 1884–1906; the Church Record of the Toledo Circuit, 1889–1915; the Church Record of the Salkum and Ferry Circuit, 1894–1938, Winlock United Methodist Church, Winlock, Wash. Records from the Winlock Baptist Church (now Cowlitz Prairie Baptist Church) and the Toledo Presbyterian Church were also available, but much less thorough. Although several Catholic churches and Mormon stations are listed in

NOTES TO CHAPTER TWO   207

area phone books, I was never able to make contact with anyone at them, despite
repeated efforts.

46. "Patchwork Potluck: A Collection of Recipes from St. Paul's Lutheran
Church, American Lutheran Church Women" (St. Paul's Lutheran Church, Winlock, Wash., 1983).

47. See, for example, *Chehalis Nugget*, April 14, 1893; *People's Advocate*, March 15,
1895, January 10, 1896.

48. For overviews of fraternal history, see Mary Ann Clawson, *Constructing Brotherhood: Class, Gender, and Fraternalism* (Princeton: Princeton University Press, 1989);
Mark C. Carnes, *Secret Ritual and Manhood in Victorian America* (New Haven: Yale
University Press, 1989); Noel P. Gist, "Secret Societies: A Cultural Study of Fraternalism in the United States," *Missouri University Studies* 15, no. 4 (1940): 5–176.

49. Mossyrock Grange, *Memories*, 157–59.

50. Sixtieth-anniversary history, undated, Montrose Rebekah Lodge #46, IOOF,
Toledo, Wash.

51. Minute Book, June 6, 1895, November 18, 1897, Montrose Rebekah Lodge
#46, IOOF, Toledo; Roll Book, 1901–18, Toledo Masons.

52. These figures are based on all of the known lodge members whose names
could be found in the 1900 manuscript census. Organizational records were available and opened to me for the Masons, Eastern Star, and Rebekahs of Toledo, and
the Eastern Star of Winlock. Members' names for other lodges were drawn largely
from newspapers which usually only listed officers or those active in planning a
particular event. Financial data was added from county tax records. According to
these figures, Masons were far more likely than Oddfellows to be farmers or laborers,
while Oddfellows were two and a half times as likely to be involved in business or
professional occupations. These figures might well be skewed, however. Farmers and
laborers were less likely to be among the leaders whose names would appear in the
newspaper. The fact that seventy-one percent of female Rebekahs lived on a farm
(compared to thirty-seven percent of Eastern Star members) also suggests that farmers, at least, were undercounted among Oddfellows.

53. Mark C. Carnes, *Secret Ritual*, 14, 31, 52.

54. Clawson, *Constructing Brotherhood*, 17, 187, 195; Mary Ann Clawson, "Nineteenth-Century Women's Auxiliaries and Fraternal Orders," *Signs: Journal of Women
in Culture and Society* 12 (1986): 41, 46–47.

55. Roll Book, 1901–18, Toledo Masons, Toledo, Wash.; Mossyrock Grange, *Memories*, 157. Klaus Bezemer of Ferry was the Lewis County Populist chair in 1892 and
William VanWoert served as a delegate to Republican conventions throughout the
1890s.

56. Minute Book, December 5, 1895, Montrose Rebekah Lodge, Toledo, Wash.

57. Clawson, *Constructing Brotherhood*, 88; Elizabeth Ann Jameson, "High-Grade
and Fissures: A Working-Class History of the Cripple Creek, Colorado, Gold Mining
District, 1890–1905" (Ph.D. diss., University of Michigan, 1987), 185; Carnes, *Secret
Ritual*, 26–32. In the Kansas towns he studied, McNall found a wide mix of occupations in the Masons, but few farmers. See McNall, *Road to Rebellion*, 119, 137.

58. *Chehalis Bee*, February 12, 1897.

59. Nix and Nix, eds., *History of Lewis County*, 40.

60. In examining late-nineteenth-century Kansas and Colorado, Scott McNall

and Richard Hogan also see some fluidity, but do draw clear class distinctions. McNall, *Road to Rebellion*, 140–41; Richard Hogan, *Class and Community in Frontier Colorado* (Lawrence: University Press of Kansas, 1990). Hogan classifies farmers as artisans.

61. *People's Advocate*, June 4, 1897. Census instructions indicated that people of mixed white and Indian ancestry living among the general population should be counted as Indians only if enrolled in a tribe or "regarded as Indians," and if they had at least one-quarter Indian blood. Bureau of the Census, *200 Years of U.S. Census Taking*, 36, 41; Hyman Alterman, *Counting People: The Census in History* (New York: Harcourt, Brace & World, 1969), 291–94.

62. *Bee Nugget*, December 5, 1912, May 10, 1918.

63. U.S. Census manuscripts for Veness precinct, 1900 and 1910; R. L. Polk & Co., *Directory of Lewis County 1922–23*, (Seattle: R. L. Polk & Co., 1923), Winlock city listings; Betsy Browne, "Blacks in Winlock," in "Stories on Winlock," Lewis County Historical Society, Chehalis, Wash.

64. *People's Advocate*, January 20, 1893.

65. *Chehalis Nugget*, March 2, 1894.

66. Ibid., January 20, 1893; *Chehalis Bee*, August 14, 1891.

67. Carlos A. Schwantes, *Radical Heritage: Labor, Socialism, and Reform in Washington and British Columbia, 1885–1917* (Vancouver: Douglas & McIntyre, 1979), 23–25.

68. *People's Advocate*, June 25, 1897.

69. *Bee Nugget*, January 4, 1901.

70. *People's Advocate*, May 20, 1898. For additional positive expressions toward Catholics, see *Chehalis Bee*, June 29, 1894, February 26, 1897; *Chehalis Nugget*, June 29, 1894. Protestant missionaries to Oregon territory in the 1840s, on the other hand, were openly, even viciously, anti-Catholic, as documented in Clark, *Eden Seekers*, 196–240; Schwantes, *Pacific Northwest*, 76–77, 81–84.

71. Register, "Chronik," 1891–; "Patchwork Potluck," St. Paul's Evangelical Lutheran Church, Winlock, Wash.

72. *People's Advocate*, February 12, 1897.

73. Church Record of the Toledo Circuit, 1889–1915; Church Record of the Winlock Methodist Episcopal Church, 1884–1906, Winlock United Methodist Church, Winlock, Wash.

74. Minute Book, August 7, 1906, March 3, 1908, Toledo Masons.

75. *Chehalis Bee*, February 13, 1891, February 2, 1894; *People's Advocate*, February 17, 1893, March 3, 1893.

76. Mossyrock Grange, *Memories*, 159; Polk, *Directory of Lewis County, 1904*, 99, 107.

77. For discussion of the integrative effects of community institutions, see Don Harrison Doyle, *The Social Order of a Frontier Community: Jacksonville, Illinois, 1825–1870* (Urbana: University of Illinois Press, 1978).

3. New Visions: Political Culture in the Farmers' Alliance

1. *Chehalis Nugget*, June 12, 1891, July 3, 1891, July 17, 1891, August 28, 1891; *Chehalis Bee*, July 3, 1891.

2. Lawrence Goodwyn, *Democratic Promise: The Populist Moment in America* (New York: Oxford University Press, 1976) xi–xii; Lawrence Goodwyn, "The Cooperative Commonwealth and Other Abstractions: In Search of a Democratic Promise," *Marxist Perspectives* 3 (Summer 1980): 8–42. Sara M. Evans and Harry C. Boyte use much the same terminology and rely heavily on Goodwyn's analysis in *Free Spaces: The Sources of Democratic Change in America* (New York: Harper & Row, 1986).

3. For example, see Michael Schwartz, *Radical Protest and Social Structure: The Southern Farmers' Alliance and Cotton Tenancy, 1880–1890* (New York: Academic Press, 1976), 110; Scott G. McNall, *The Road to Rebellion: Class Formation and Kansas Populism, 1865–1900* (Chicago: University of Chicago Press, 1988), 209. For a more inclusive image of the midwestern Alliance, see Maryjo Wagner, "Farms, Families, and Reform: Women in the Farmers' Alliance and Populist Party" (Ph.D. diss., University of Oregon, 1986); Jane Taylor Nelsen, ed., *A Prairie Populist: The Memoirs of Luna Kellie* (Iowa City: University of Iowa Press, 1992), 135–45.

4. St. Helens Club Minute Book, Lewis County Historical Society, Chehalis, Wash.; *Chehalis Bee*, January 12, 1894, October 5, 1894; *Bee Nugget*, April 15, 1910, April 22, 1910, February 24, 1911, March 31, 1911, January 30, 1913, March 21, 1913, July 18, 1913, January 24, 1914, January 10, 1919; *Winlock News*, January 13, 1922, June 16, 1922.

5. Maryjo Wagner argues that women's involvement in the Alliance and People's party was a natural extension of their partnership with men on the farm but gives little attention to their role in other aspects of community life. See Wagner, "Farms, Families, and Reform," 1–18. Among the few other works that specifically address women's involvement in the Farmers' Alliance and Populist party are: Nelsen, *A Prairie Populist;* Julie Roy Jeffrey, "Women in the Southern Farmers' Alliance: A Reconsideration of the Role and Status of Women in the Late Nineteenth Century South," *Feminist Studies* 3 (Fall 1975): 72–90; Robert C. McMath, Jr., *Populist Vanguard: A History of the Southern Farmers' Alliance* (Chapel Hill: University of North Carolina Press, 1975); Pauline Adams and Emma S. Thornton, *A Populist Assault: Sarah E. Van de Vort Emery on American Democracy, 1862–95* (Bowling Green, Ohio: Bowling Green State University Popular Press, 1982); Annie L. Diggs, "The Women in the Alliance Movement," *The Arena* 6 (July 1892): 160–79.

6. Thomas A. Woods, *Knights of the Plow: Oliver H. Kelley and the Origins of the Grange in Republican Ideology* (Ames: Iowa State University, 1991), xv–xxii, 178; Harriet Ann Crawford, *The Washington State Grange, 1889–1924: A Romance of Democracy* (Portland, Oreg.: Binfords & Mort, 1940), 13. The origins and early history of the Grange are discussed in Solon J. Buck, *The Granger Movement: A Study of Agricultural Organization and Its Political, Economic and Social Manifestation* (Cambridge: Harvard University Press, 1933); James Dabney McCabe, *History of the Grange Movement: Or the Farmers' War Against Monopolies* by Edward Winslow Martin [pseud.], (New York: Augustus M. Kelley, 1969 [1873]); Charles M. Gardner, *The Grange—Friend of the Farmer: A Concise Reference History of America's Oldest Farm Organization, and the Only Rural Fraternity in the World, 1867–1947* (Washington, D.C.: The National Grange, 1949); O. H. Kelley, *Origin and Progress of the Order of the Patrons of Husbandry in the United States; A History from 1866 to 1873* (Philadelphia: J.A. Wagenseller, 1875).

7. For background on the Farmers' Alliance and genesis of the People's party, see Robert C. McMath, Jr., *American Populism: A Social History, 1877–1898* (New York:

Hill & Wang, 1993); Donna A. Barnes, *Farmers in Rebellion: The Rise and Fall of the Southern Farmers' Alliance and People's Party in Texas* (Austin: University of Texas, 1984); Goodwyn, *Democratic Promise*; Steven Hahn, *The Roots of Southern Populism: Yeoman Farmers and the Transformation of the Georgia Upcountry, 1850–1890* (New York: Oxford, 1983); John D. Hicks, *The Populist Revolt: A History of the Farmers' Alliance and the People's Party* (Lincoln: University of Nebraska Press, 1961 [1931]).

8. Fred R. Yoder, "The Farmers' Alliance in Washington—Prelude to Populism," *Research Studies of the State College of Washington* 16 (1948); Crawford, *Washington State Grange*, 12–16; Max Gerhart Geier, "A Comparative History of Rural Community on the Northwest Plains: Lincoln County, Washington and the Wheatland Region, Alberta, 1880–1930," (Ph.D. diss., Washington State University, 1990), 97, 147–49; Carlos A. Schwantes, *Radical Heritage: Labor, Socialism, and Reform in Washington and British Columbia* (Vancouver, B.C.: Douglas & McIntyre, 1979), 54–55.

9. *People's Advocate*, December 2, 1892, August 2, 1894; *Chehalis Nugget*, June 12, 1891; Yoder, "The Farmers' Alliance," 130–31.

10. Studies of farmer organizations in other areas have shown that the more prosperous often assumed leadership roles. McMath, *Populist Vanguard*, chap. 5; Hicks, *Populist Revolt*, 52; Goodwyn, *Democratic Promise*, xx; Hahn, *Roots of Southern Populism*, 165.

11. Most studies of Washington Populism focus on the eastern part of the state where rates of indebtedness were high and a single-crop economy predominated, relying on these factors to explain the movement's appeal. See Yoder, "Farmers' Alliance in Washington," 136–45; Schwantes, *Radical Heritage*, 54–57. Schwantes emphasizes the urban-labor rather than the rural roots of Populism in his general history of the Northwest, and barely mentions the Farmers' Alliance. See Carlos A. Schwantes, *The Pacific Northwest: An Interpretive History* (Lincoln: University of Nebraska Press, 1989), 268–70. Studies that have found that middle-aged, well-established farmers were most likely to join reform movements include McMath, *Populist Vanguard*, chap. 5; John David Dibbern, "Grass Roots Populism: Politics and Social Structure in a Frontier Community" (Ph.D. diss., Stanford University, 1980), 57–69; Kathleen Diane Moum, "Harvest of Discontent: The Social Origins of the Nonpartisan League, 1880–1922" (Ph.D. diss., University of California, Irvine, 1986), 14.

12. *People's Advocate*, December 23, 1892. For other reports of hard times and economic recovery at the end of the decade, see *People's Advocate*, July 20, 1894, August 24, 1894, and July 18, 1899.

13. Ibid., December 29, 1893. The last mention of the Telephone Alliance in the *People's Advocate* was on June 4, 1897. The sources give no indication of how the local got its name.

14. *People's Advocate*, March 10, 1893, April 14, 1893, September 3, 1893; *Chehalis Nugget*, September 29, 1893.

15. *People's Advocate*, April 14, 1893, September 3, 1893, September 29, 1893, August 21, 1896; see also *Chehalis Bee*, April 6, 1894, May 11, 1894; *Chehalis Nugget*, December 22, 1893.

16. *Chehalis Nugget*, September 11, 1891.

17. *Chehalis Bee*, July 13, 1894, June 4, 1897.

18. *Chehalis Bee*, June 24, 1892, July 1, 1892.

19. J. Alexander to John Alexander, July 12, 1892, (W. A. Alexander Collection, Washington Historical Society, Tacoma, Wash.).

20. *People's Advocate,* June 16, 1893, July 14, 1893, June 21, 1895; *Chehalis Bee,* July 7, 1893.

21. For discussion of gender implications of public July Fourth celebrations earlier in the century, see Mary Ryan, *Women in Public: Between Banners and Ballots, 1825–1880* (Baltimore: Johns Hopkins University Press, 1990) 22–35; Richard B. Scott, *Workers in the Metropolis: Class, Ethnicity, and Youth in Antebellum New York City* (Ithaca: Cornell University Press, 1990).

22. *People's Advocate,* January 13, 1893, July 13, 1894, November 1, 1895, January 10, 1896.

23. Ibid., December 9, 1892, March 17, 1893, September 3, 1893, October 20, 1893, December 15, 1893.

24. Wagner, "Farms, Families, and Reform," 33–47; Nelsen, *A Prairie Populist;* Donald B. Marti, *Women of the Grange: Mutuality and Sisterhood in Rural America, 1866–1920* (New York: Greenwood, 1991) 27–29.

25. *People's Advocate,* July 3, 1891, January 13, 1893, January 12, 1894, November 1, 1895.

26. Ibid., February 24, 1893.

27. Ibid., March 1, 1895, March 8, 1895.

28. For other examples of "especially the ladies" in Alliance announcements, see McNall, *Road to Rebellion,* 231; Marilyn P. Watkins, "The Populist Movement in Hartford, Michigan: Politics in Community Life," (University of Michigan, 1984).

29. *People's Advocate,* November 9, 1894.

30. Ibid., August 9, 1895.

31. Ibid., May 24, 1895.

32. Ibid., November 9, 1894.

33. Ibid., May 10, 1895.

34. Studies of the Alliance in the South and in Kansas suggest exclusion of women from Alliance business meetings was routine in those locales. Schwartz, *Radical Protest,* 110; McNall, *The Road to Rebellion,* 209.

35. *People's Advocate,* August 23, 1895, November 8, 1895.

36. See Ibid., October 14, 1892 for quotation, and July 7, 1893 for reference to "former Republican." See also ibid., December 30, 1892; and *Chehalis Bee,* June 8, 1894.

37. *People's Advocate,* March 10, 1893.

38. Ibid., January 27, 1893, April 8, 1892; *Chehalis Bee,* May 8, 1896.

39. *People's Advocate,* April 26, 1895, June 8, 1894.

40. Ibid., December 2, 1892. See also Mitchell, *Political Education,* for a fuller discussion of education in the Alliance, although he gives little attention to gender issues or women.

41. *People's Advocate,* June 22, 1894; *Chehalis Bee,* June 8, 1894.

42. *People's Advocate,* February 16, 1894, March 2, 1894, June 1, 1894.

43. Ibid., June 1, 1894.

44. Ibid., March 9, 1894, October 2, 1896.

45. *Chehalis Bee,* October 23, 1896.

46. *People's Advocate,* March 10, 1893, March 31, 1893.

47. Ibid., March 1, 1895; *Chehalis Bee,* April 20, 1894.

4. Populists and Republicans: National Parties and Local Issues

1. *Chehalis Bee,* May 20, 1892, August 12, 1892; *People's Advocate,* November 11, 1892.

2. Edward Bellamy's *Looking Backward* (1888) and Henry George's *Progress and Poverty* (1879) were best sellers in the late nineteenth century. In Bellamy's book, the hero awakens from a hundred-year sleep to find himself in a transformed society, which has overcome the class conflict and inequities of the nineteenth century and allows all to live in happy comfort. George proposed ending vast disparities in wealth and opportunity through a Single Tax on land that would discourage speculation and open opportunity to many. Although the Populists adopted neither author's specific proposals, many people came to the reform movement after reading their indictments of the industrial economy.

3. The Populist platform has been widely reprinted. A good concise history of the Alliance movement and the creation of the Populist party is Robert C. McMath, Jr., *American Populism: A Social History, 1877–1898* (New York: Hill & Wang, 1993).

4. A number of master's and doctoral theses have been written that focus on state-level politics during Washington's Populist period, including Robert Donald Saltvig, "The Progressive Movement in Washington" (Ph.D. diss., University of Washington, 1966); Stephen H. Peters, "The Populists and the Washington Legislature, 1893–1900" (master's thesis, University of Washington, 1967); Margaret H. Thompson, "The Writings of John Rankin Rogers" (master's thesis, University of Washington, 1947). See also Thomas Wayne Riddle, *The Old Radicalism: John R. Rogers and the Populist Movement in Washington* (New York: Garland, 1991).

5. *People's Advocate,* October 7, 1892.

6. Ibid., November 2, 1894, for quotation. See also March 6, 1896.

7. Ibid., August 2, 1895.

8. Ibid., January 22, 1897 (quotation), January 29, 1897; January 18, 1895 for "reprehensible" reference.

9. Ibid., February 12, 1897.

10. *Chehalis Bee,* May 8, 1896.

11. Ibid., May 1, 1896; *People's Advocate,* February 18, 1898, May 6, 1898, May 27, 1898.

12. *Winlock Pilot* quoted in *Chehalis Bee,* September 3, 1897; see *Chehalis Bee,* August 20, 1897 for *Bee* correspondent quotation.

13. *People's Advocate,* January 14, 1898. Reports of booming mills can be found in *Chehalis Bee,* September 24, 1897, November 5, 1897; *People's Advocate,* July 16, 1897, August 6, 1897, October 1, 1897.

14. *People's Advocate,* July 28, 1899.

15. *Chehalis Nugget,* August 5, 1892; *Chehalis Bee,* May 20, 1892, August 26, 1892, September 16, 1892.

16. *Chehalis Bee,* September 16, 1892; *Chehalis Nugget,* September 26, 1890, September 28, 1894; *People's Advocate,* October 21, 1892.

17. *People's Advocate*, March 3, 1893.

18. Ibid., February 16, 1894.

19. Ibid., May 25, 1894, February 25, 1898.

20. *Chehalis Bee*, August 24, 1894; *People's Advocate*, August 2, 1894, August 10, 1894.

21. *People's Advocate*, February 25, 1898; *Chehalis Bee*, May 10, 1895.

22. *People's Advocate*, June 7, 1895.

23. *Chehalis Bee*, January 25, 1895; *People's Advocate*, June 17, 1898. Anna Wuestney vs. E. H. Painter et al., #1398 Superior Court of Washington for Lewis County, Lewis County Clerk's Office, Chehalis, Washington. No record of the final disposition of the case appears in court records, indicating the suit was either settled out of court or dropped.

24. *Chehalis Bee*, January 25, 1895.

25. Donald L. Kinzer, *An Episode in Anti-Catholicism: The American Protective Association* (Seattle: University of Washington Press, 1964), 45, 106–11; John Higham, *Strangers in the Land: Patterns of American Nativism 1860–1925* (New York: Atheneum, 1974 [1955]), 53–67, 68–74, 80–86; David H. Bennett, *The Party of Fear: From Nativist Movements to the New Right in American History* (Chapel Hill: University of North Carolina Press, 1988), 163–66, 171–78.

26. For positive references to local Catholic immigrants, see *Chehalis Bee*, June 29, 1894, February 26, 1897; *Chehalis Nugget*, June 29, 1894. For examples of anti-Chinese sentiment, see *Chehalis Nugget*, January 20, 1893, *Chehalis Bee*, August 14, 1891.

27. *Chehalis Nugget*, November 2, 1894, July 5, 1895.

28. *Chehalis Bee*, May 8, 1896; *People's Advocate*, June 19, 1896.

29. *Chehalis Bee*, September 4, 1896.

30. Ibid., September 11, 1896; *People's Advocate*, November 13, 1896.

31. *Chehalis Bee*, November 13, 1896.

32. Kinzer, *Episode*, 219; Higham, *Strangers*, 104; Paul Kleppner, *The Cross of Culture: A Social Analysis of Midwestern Politics 1850–1900* (New York: Free Press, 1970), 267. Stanley Parsons also found that APA influence was strongest in the Republican party and that it was rejected by the Populists in rural Nebraska. Stanley B. Parsons, *The Populist Context: Rural vs. Urban Power on a Great Plains Frontier* (Westport, Conn.: Greenwood, 1973), 107–10.

33. Enough ink has already been spilt debating whether the Populists in general were more or less nativist than the bulk of the U.S. population at the time. Even McMath, an otherwise sympathetic historian, has accepted the assertion of Schwantes that Washington Populism was rooted primarily in the urban labor movement and had a distinctively Sinophobic cast. This reading of northwest Populism is clearly not supported by the evidence from rural areas. See McMath, *American Populism*, 120–22; Carlos A. Schwantes, *The Pacific Northwest: An Interpretive History* (Lincoln: University of Nebraska Press, 1989), 268–70.

34. On average, 96 percent of households in the eastern precincts were headed by farmers, 33.5 percent of adults were assessed real property taxes, and 47 percent remained to be counted in the 1910 census. In comparison, 51 percent of the Winlock area's households were farmer-headed, 20 percent of adults were assessed real property tax, and 34 percent were counted in both censuses.

35. For complete documentation of election returns and precinct demographic characteristics, see Marilyn P. Watkins, "Political Culture and Gender in Rural Community Life: Agrarian Activism in Lewis County, Washington, 1890–1925" (Ph.D. diss., University of Michigan, 1991). Jeffrey Ostler has also noted that economic characteristics alone did not determine the level of Populist strength, arguing that the degree of competitiveness between the two major parties strongly influenced farmers' willingness to turn to a third party. Jeffrey Ostler, "Why the Populist Party Was Strong in Kansas and Nebraska but Weak in Iowa," *Western Historical Quarterly* 23 (November 1992): 452–74. In contrast to the findings here, many historians have found ethnicity and religion to be major factors in voter affiliations. See Kleppner, *Cross of Culture*; Lowell J. Soike, *Norwegian Americans and the Politics of Dissent 1880–1924* (Northfield, Minn.: Norwegian-American Historical Association, 1991); Richard Jensen, *The Winning of the Midwest: Social and Political Conflict 1888–1896* (Chicago: University of Chicago Press, 1971); Frederick Luebke, *Immigrants and Politics: The Germans of Nebraska, 1880–1900* (Lincoln: University of Nebraska Press, 1969).

36. *Chehalis Nugget*, February 6, 1891.

37. *Chehalis Bee*, March 11, 1892; see also *Chehalis Nugget*, July 17, 1891.

38. *Chehalis Bee*, April 15, 1892.

39. Ibid., May 13, 1892, May 27, 1892, June 3, 1892.

40. Ibid., June 3, 1892, July 8, 1892, August 5, 1892. Of the twenty-three delegates from the two study areas whose affiliations could be identified, only two were Alliance members; six others later became Populists, two were Democrats, and the other thirteen were Republicans.

41. *People's Advocate*, December 23, 1892, December 30, 1892; *Chehalis Nugget*, January 6, 1893, January 27, 1893; *Chehalis Bee*, January 6, 1893, January 13, 1893.

42. The new bill continued those provisions of the much maligned 1890 bill that had escaped general criticism, including a poll tax on all men between the ages of twenty-one and fifty living outside incorporated areas, and town-meeting-style elections of road district supervisors. It also explicitly allowed anyone who wished to pay their taxes in labor to do so. Both the poll tax and the road tax assessed on property could be paid off at the rate of $2.00 for an eight-hour day of "diligent" work under the road supervisor's direction, or $4.00 per day if horses and equipment were also provided. The supervisor himself received $2.50 for each day worked on the roads. In 1901, provisions for paying road taxes through labor were replaced with a statement that taxes were "payable in money without any exemption whatsoever," and major work was now to be put out to bid rather than being performed by local men. In 1913 the poll tax was eliminated. *Session Laws of Washington State, 1890*, 145–49; *1893*, chap. LXIX; *1901*, H.B. No. 475; *1913*, chap. 151.

43. *Chehalis Bee*, June 21, 1895, March 13, 1896; *People's Advocate*, May 19, 1899, March 31, 1899.

5. Progressive Populists: The Grange in Lewis County

1. Henry McQuigg, Jr., "Recollections," Ethel Grange, Ethel, Wash., n.d.; "Lewis County," list of subordinate granges, n.d., Washington State Grange, Olympia, Wash.

2. Harriet Ann Crawford, *The Washington State Grange, 1889–1924: A Romance of Democracy* (Portland, Oreg.: Binfords & Mort, 1940), 110.

3. Goodwyn describes the Farmers' Alliance and particularly its cooperative endeavors as a similar kind of "free social space." Lawrence Goodwyn, *Democratic Promise: The Populist Moment in America* (New York: Oxford University Press, 1976), xi–xii; Lawrence Goodwyn, "The Cooperative Commonwealth and Other Abstractions: In Search of a Democratic Promise," *Marxist Perspectives* 3 (Summer 1980): 8–42. Evans and Boyte follow Goodwyn's analysis for the Alliance and also find evidence of free space in the labor, women's, and civil rights movements. See Sara M. Evans and Harry C. Boyte, *Free Spaces: The Sources of Democratic Change in America* (New York: Harper & Row, 1986). Edward P. Morgan uses different terminology, but also states that democracy is rooted in small-scale community connected to the larger society through political activism. Edward P. Morgan, *The Sixties Experience: Hard Lessons about Modern America* (Philadelphia: Temple University Press, 1991), 8, 274, 282.

4. While the central role of urban middle-class women in creating and lobbying for the Progressive agenda has been well documented by historians, the activities of farm women have been largely hidden behind the male leadership of their organizations. For a sampling of discussion of urban women in Progressive reform, see Robyn Muncy, *Creating a Female Dominion in American Reform 1890–1935* (New York: Oxford University Press, 1991), 27–30; Peggy Pascoe, *Relations of Rescue: The Search for Female Moral Authority in the West, 1874–1939* (New York: Oxford University Press, 1990), 178–83; Anne Firor Scott, *Natural Allies: Women's Associations in American History* (Urbana: University of Illinois Press, 1991), 1–5.

5. Crawford, *Washington State Grange,* 12–16; Gus Norwood, *Washington Grangers Celebrate a Century* (Seattle: Washington State Grange, 1988), 55–58.

6. *Proceedings of the Washington State Grange, 1891,* 28.

7. Quotation from *Proceedings of the Grange, 1894,* 32; see also *1893,* 31, 35, 65; *1896,* 10.

8. Crawford, *Washington State Grange,* 189–90, 200. According to the records of the Washington State Grange, nineteen subordinate Granges were organized in Lewis county between 1903 and 1910, and fourteen more between 1913 and 1920. Several of those lasted only a few years, and seven went under during the politically divisive early 1920s, but fourteen of those early Granges flourished into the 1980s and several new locals were instituted after 1920.

9. Membership rolls were available from several of the Granges, including Alpha, Ethel, and Silver Creek, all of which began in 1904, and Cowlitz and Cougar Flat, which began in 1920. Hope (founded 1904) and St. Urban (founded 1917) had minutes and other records which yielded many members' names. For Granges which had folded prior to this study, including Eden Prairie, Napavine, and Pumpkin Center near Winlock, only scattered newspaper references were available. Altogether, 361 Grange members were linked to their 1910 census record.

10. *Chehalis Nugget,* June 22, 1894; "Alpha Grange History," 2, Alpha Grange.

11. *The Toledo Community Story, 1853–1953,* 38, 121, Seattle Public Library. *Proceedings of the Grange, 1904,* 55; *1907,* 2; *1908,* 2; *1910,* 2.

12. These figures and the ones that follow are based on those people known to have joined the Grange between 1903 and 1921 whose names could be located in the 1910 U.S. Census manuscripts for the study precincts. Property valuations were

obtained from Lewis county assessor figures for 1909, now housed in the Washington State Archives, Olympia. Because Grange records were incomplete and name matching sometimes difficult, figures should be interpreted cautiously.

13. In comparison, McMath and McNall found women to constitute about one-fourth of Farmers' Alliance membership. Robert C. McMath, Jr., *Populist Vanguard: A History of the Southern Farmers' Alliance* (Chapel Hill: University of North Carolina Press, 1975), chap. 5; Scott G. McNall, *The Road to Rebellion: Class Formation and Kansas Populism, 1865–1900* (Chicago: University of Chicago Press, 1988), 238.

14. Carlos A. Schwantes, "Farmer-Labor Insurgency in Washington State: William Bouck, the Grange, and the Western Progressive Farmers," *Pacific Northwest Quarterly* 76 (January 1985): 8. Donald Marti notes that in several states, including Pennsylvania and New Hampshire, Granges had many nonfarmer members. Donald B. Marti, *Women of the Grange: Mutuality and Sisterhood in Rural America, 1866–1920* (New York: Greenwood, 1991), 2–4.

15. *Proceedings of the Grange, 1905*, 51.

16. The following Grange records were used for this study: Alpha Grange, Minute Books 1917–25, Roll Book 1904–28; Cougar Flat Grange, Minute Books 1922–27, Roll Book 1920–, Dues Book 1920–; Ethel Grange, Minute Books 1919–30, Roll Book 1904–38, Dues Book 1904–25; Hope Grange, Minute Books 1911–25, Dues Books 1904–33; Silver Creek Grange, Minute Books 1904–11 and 1915–30, Roll Book 1904–23, Dues Books 1905–35; St. Urban Grange, Minute Books 1917–20 and 1925–29. Current Grange officers are responsible for preserving the records.

17. Minute Book, June 3, 1916, November 10, 1916, February 20, 1915, Silver Creek Grange; Minute Book, February 24, 1921, Alpha Grange; Minute Book, March 17, 1923, Cougar Flat Grange.

18. Minute Book, July 5, 1913, Hope Grange; Minute Book, November 6, 1915, November 2, 1910, November 4, 1905, Silver Creek Grange.

19. Minute Book, October 30, 1916, Silver Creek Grange; Minute Book, September 23, 1920, May 12, 1921, Cowlitz Prairie Grange; Minute Book, August 17, 1912, May 4, 1918, Hope Grange.

20. Alpha Grange History, undated, 6; Minute Book, March 5, 1921, Ethel Grange; Minute Book, January 18, 1919, February 15, 1919, St. Urban Grange; Minute Book, June 21, 1913, July 5, 1913, Hope Grange; Minute Book, December 4, 1915, Silver Creek Grange.

21. See for example, Minute Book, June 28, 1917, May 10, 1919, March 13, 1920, April 21, 1923, January 5, 1924, Alpha Grange.

22. *Bee Nugget*, February 11, 1910, May 13, 1910, February 10, 1911, February 15, 1912, May 9, 1913, November 14, 1913, February 13, 1914.

23. Ibid., January 21, 1910, January 27, 1911.

24. Ibid., January 21, 1910.

25. For the suffrage campaign in the Northwest, see T. A. Larson, "The Woman Suffrage Movement in Washington," *Pacific Northwest Quarterly* 67 (April 1976): 49–62; Ruth Barnes Moynihan, *Rebel for Rights: Abigail Scott Duniway* (New Haven: Yale University Press, 1983); Ruth Barnes Moynihan, "Of Women's Rights and Freedom: Abigail Scott Duniway," in Karen J. Blair, ed., *Women in Pacific Northwest History: An Anthology* (Seattle: University of Washington Press, 1988), 9–24; Patricia Voeller Hor-

ner, "May Arkwright Hutton: Suffragist and Politician," in Blair, ed., *Women in Pacific Northwest History,* 25–42; Lauren Kessler, "The Fight for Woman Suffrage and the Oregon Press," in Blair, ed., *Women in Pacific Northwest History,* 43–58; G. Thomas Edwards, *Sowing Good Seeds: The Northwest Suffrage Campaigns of Susan B. Anthony* (Portland: Oregon Historical Society Press, 1990). See also Eleanor Flexnor, *Century of Struggle: The Woman's Rights Movement in the United States* (Cambridge: Belknap Press of Harvard University Press, 1959); Carolyn Stefanco, "Networking on The Frontier: The Colorado Women's Suffrage Movement, 1876–1893," in Susan Armitage and Elizabeth Jameson, eds., *The Women's West* (Norman: University of Oklahoma Press, 1987), 265–76.

26. For discussion of women's involvement in the more politically conservative New York Grange see Paula Baker, *The Moral Frameworks of Public Life: Gender, Politics, and the State in Rural New York, 1870–1930* (New York: Oxford University Press, 1991), 59–64; Nancy Grey Osterud, *Bonds of Community: The Lives of Farm Women in Nineteenth-Century New York* (Ithaca: Cornell University Press, 1991), 255–62. Osterud found a similar expansion of opportunities. Joan M. Jensen and Gloria Ricci Lothrop briefly discuss the early endorsement of women's rights by the California Grange in *California Women: A History* (San Francisco: Boyd and Fraser, 1987), 23. Marti, *Women of the Grange,* documents the tension in the Grange nationally between belief in gender equality and domesticity, but my interpretation of women's political role differs significantly from his. Marti focuses his political discussions almost exclusively on "women's" issues such as suffrage and emphasizes the differences in men's and women's educational interests, in practical farming and domestic science, respectively. Such a split was not evident in the Lewis county Granges.

27. James Daubney McCabe, *History of the Grange Movement: Or the Farmers' War Against Monopolies* by Edward Winslow Martin [pseud.] (New York: Augustus M. Kelley, 1969 [1873]), 458.

28. McCabe, *History of the Grange,* 456; see also 420, 451.

29. McCabe, *History of the Grange,* 428.

30. Donald B. Marti, "Sisters of the Grange: Rural Feminism in the Late Nineteenth Century," 58 *Agricultural History* (July 1984): 247–61.

31. *Proceedings of the Grange, 1902,* 35; *1891,* 25–26; *1899,* 30.

32. *Proceedings of the Grange, 1893,* 66–67; *1894,* 54.

33. Quotation in *Proceedings of the Grange, 1892,* 78; see also *1893,* 41, 61.

34. *Proceedings of the Grange, 1891,* 51.

35. For continued strong endorsements of women's suffrage, see *Proceedings of the Grange, 1907,* 84; *1909,* 16, 30; *1910,* 31, 63. See also Marti, *Women of the Grange,* 119.

36. Quotation from *Proceedings of the Grange, 1907,* 96. See also Minute Book, November 20, 1915, Hope Grange; *Bee Nugget,* January 13, 1911; Minute Book, March 2, 1907, April 6, 1907, March 18, 1916, October 13, 1916, Silver Creek Grange.

37. In 1905, Anna Leonard was master of Sunnyside Grange in Castle Rock, Mary F. White of Pleasant Valley Grange in Colfax, and Maggie Hanlon of Cape Horn Grange; in 1906 there were also three women masters in the state, and four in 1907. See *Proceedings of the Grange, 1905,* 120–29; *1906,* 103–8; *1907,* 101–7.

38. See Minute Book, November 22, 1917, January 13, 1921, Alpha Grange; *Western Progressive Farmer*, May 20, 1923.

39. Alpha Grange History, 7–8; Minute Book, March 1, 1919, March 15, 1919, April 19, 1919, St. Urban Grange; Minute Book, January 17, 1914, March 21, 1914, November 21, 1914, November 6, 1915, Hope Grange.

40. Roll Book and Minute Book, June 15, 1912, August 17, 1912, March 15, 1912, April 5, 1913, Hope Grange; U.S. Census manuscripts for Prescott precinct, 1910.

41. Minute Book, December 1, 1906, January 5, 1907, October 5, 1907, September 7, 1917, Silver Creek Grange; Minute Book, July 26, 1917, May 8, 1920, June 24, 1920, Alpha Grange.

42. Minute Book, December 21,1907, Silver Creek Grange; Minute Book, November 22, 1917, Alpha Grange; Minute Book, December 16, 1911, December 4, 1915, November 1, 1919, Hope Grange; Officers' Roll Book, 1918, 1920–23, St. Urban Grange; *Bee Nugget*, January 14, 1910. Marti also found women lecturers and secretaries to be common in Grange locals nationally, with a sprinkling of women masters. Marti, *Women in the Grange*, 27–29.

43. See *Proceedings of the Grange, 1893–1920*.

44. Minute Book, June 5, 1920, Silver Creek Grange; Minute Book, March 20, 1920, April 3, 1920, St. Urban Grange.

45. Minute Book, November 4, 1922, November 18, 1922, Cougar Flat Grange; Minute Book, November 1, 1919, December 4, 1919, Ethel Grange.

46. Minute Book, January 17, 1920, St. Urban Grange; Ladies' Auxiliary Notes, n.d., St. Urban Grange; Minute Book, June 3, 1922, Cougar Flat Grange.

47. See, for example, Gerda Lerner, *The Grimke Sisters from South Carolina: Rebels against Slavery* (Boston: Houghton Mifflin, 1967); Ruth Bordin, *Women and Temperance: The Quest for Power and Liberty, 1873–1900* (Philadelphia: Temple University Press, 1981); Meredith Tax, *The Rising of the Women: Feminist Solidarity and Class Conflict, 1880–1917* (New York: Monthly Review Press, 1980).

48. Minute Book, June 20, 1908, Silver Creek Grange; *Proceedings of the Grange, 1910*, 31, 91, 94, 124, 133.

49. *Bee Nugget*, May 13, 1910, May 20, 1910; *Proceedings of the Grange, 1910*, 124.

50. *Bee Nugget*, August 5, 1910, September 23, 1910.

51. Minute Book, May 21, 1910, July 6, 1910, August 3, 1910, Silver Creek Grange; *Bee Nugget*, May 13, 1910, May 27, 1910, August 26, 1910, September 2, 1910, October 21, 1910, October 28, 1910.

52. Carol K. Coburn, *Life at Four Corners: Religion, Gender, and Education in a German-Lutheran Community, 1865–1945* (Lawrence: University Press of Kansas, 1992), 127–28; Fred Luebke, *Immigrants and Politics: The Germans in Nebraska, 1880–1900* (Lincoln: University of Nebraska Press, 1969), 128–32.

53. Both Coburn and Luebke state that German immigrants living in high densities and involved in immigrant churches maintained a separate identity and distinctive voting patterns, while those more dispersed tended to adopt the patterns of the dominant society. Coburn, *Life at Four Corners*, 6–8; Luebke, *Immigrants and Politics*, 53–70. The influence of "pietist" churches (most of them Protestant) vs. "liturgical" (e.g., Catholic and Episcopal) churches on voting behavior of other immigrant groups is discussed in Richard J. Jensen, *The Winning of the Midwest: Social*

and Political Conflict, 1888–96 (Chicago: University of Chicago Press, 1971), 6, 95–98; Paul Kleppner, *The Cross of Culture: A Social Analysis of Midwestern Politics, 1850–1900* (New York: Free Press, 1970), 71–73. Exception to the notion of pietist/liturgical differences affecting Norwegian-American voting behavior is taken by Lowell J. Soike, *Norwegian Americans and the Politics of Dissent 1880–1924* (Northfield, Minn.: Norwegian-American Historical Association, 1991), 35–43. No distinctive voting patterns could be detected in this study for Scandinavian immigrants, who did not build their own churches. The distinctive patterns for Finns are discussed below.

54. Election returns were gathered from Secretary of State, General Election Returns, 1910 and 1914, Washington State Archives, Olympia. See also *Bee Nugget*, April 29, 1910.

55. *Bee Nugget*, September 12, 1912, September 26, 1912, October 3, 1912; Secretary of State, General Election Returns, 1912, Washington State Archives, Olympia.

56. *Bee Nugget*, November 21, 1912.

57. Ibid., September 21, 1911, September 12, 1912.

58. Most accounts of Progressivism emphasize its urban base, even when characterizing the movement in quite different ways. For example, see Robert H. Wiebe, *The Search for Order, 1877–1920* (New York: Hill & Wang, 1967); Samuel P. Hays, "The Politics of Reform in Municipal Government in the Progressive Era," in David M. Kennedy, ed., *Progressivism: The Critical Issues* (Boston: Little, Brown, 1971) 87–108; Gabriel Kolko, "The Trimph of Conservatism," in Kennedy, ed., *Progressivism: The Critical Issues,* 130–146; James Weinstein, *The Corporate Ideal in the Liberal State: 1900–1918* (Boston: Beacon, 1968); Richard Hofstadter, *The Age of Reform: From Bryan to FDR* (New York: Vintage, 1955); Alan Dawley, *Struggles for Justice: Social Responsibility and the Liberal State* (Cambridge: Belknap Press of Harvard University Press, 1991); Olivier Zunz, *Making America Corporate 1870–1920* (Chicago: University of Chicago Press, 1990); Muncy, *Creating a Female Dominion;* Linda Gordon, "The New Feminist Scholarship on the Welfare State," in Linda Gordon, ed., *Women, the State, and Welfare* (Madison: University of Wisconsin Press, 1990), 9–35. Elizabeth Sanders, in contrast, argues that periphery farmers and their allies were the driving force behind congressional passage of Progressive legislation. Elizabeth Sanders, "Farmers and the State in the Progressive Era," in Edward S. Greenberg and Thomas F. Mayer, eds., *Changes in the State: Causes and Consequences* (Newbury Park, Calif.: Sage Publications, 1990), 183–205. For the Progressive movement in Washington, see Carlos A. Schwantes, *Radical Heritage: Labor, Socialism, and Reform in Washington and British Columbia, 1885–1917* (Vancouver, B.C.: Douglas & McIntyre, 1979), 107, 161; Robert Donald Saltvig, "The Progressive Movement in Washington" (Ph.D. diss., University of Washington, 1966), 125; Jonathan Dembo, *Unions and Politics in Washington State, 1885–1935* (New York: Garland, 1983); Charles Byler, "Austin E. Griffiths: Seattle Progressive Reformer," *Pacific Northwest Quarterly* 76 (January 1985): 22–32.

59. Saltvig, "Progressivism," 128–29, 178; Norwood, *Washington Grangers*, 70.

60. Saltvig, "Progressivism," 175.

61. See ibid.; Barbara Winslow, "The Decline of Socialism in Washington: 1910–1925" (master's thesis, University of Washington, 1969); Hamilton Cravens, "A His-

tory of the Washington Farmer-Labor Party, 1918–1924" (master's thesis, University of Washington, 1962); Norman Frank Tjaden, "Populists and Progressives of Washington: A Comparative Study" (master's thesis, University of Washington, 1960); Dembo, *Unions and Politics.*

62. Saltvig, "Progressivism," 349–51; Crawford, *Grange,* 175–78; Cravens, "Farmer-Labor Party," 25–29.

63. Descriptions of these events can be found in Crawford, *Grange,* 182–88; Saltvig, "Progressivism," 285–91, 331–35; Jonathan Dembo, *Unions and Politics,* 69–184.

64. Minute Book, September 2, 1916, Hope Grange.

65. Alpha Grange History, 2, 7; Minute Book, May 2, 1914, August 1, 1914, Hope Grange; *Bee Nugget,* March 7, 1913.

66. *Bee Nugget,* July 22, 1910, September 2, 1910, May 29, 1914; Secretary of State, General Election Returns, 1912, Washington State Archives, Olympia. In Lewis county the recall passed 2,828 to 1,230, and the initiative and referendum 2,672 to 1,228.

67. Napavine and the surrounding countryside had finally been separated with the town's incorporation. The new rural precinct was named Emery.

68. *Bee Nugget,* February 22, 1901.

69. Ibid., February 28, 1913, February 25, 1910, April 8, 1910, April 29, 1910, July 29, 1910, September 9, 1910, September 5, 1912.

70. Ibid., December 9, 1910.

71. Ibid., November 9, 1911.

72. Minute Book, June 30, 1907, November 7, 1919, April 17, 1920, May 7, 1921, December 15, 1923, Silver Creek Grange.

73. Minute Book, January 6, 1923, Cougar Flat Grange; Minute Book, August 7, 1906, March 3, 1908, Toledo Masons.

74. *Bee Nugget,* August 8, 1913, May 8, 1914.

75. *Appeal to Reason,* May 11, 1912.

76. David Sarasohn, *The Party of Reform: Democrats in the Progressive Era* (Jackson: University of Mississippi Press, 1989), x–xv, 119.

77. The extent to which early twentieth-century Socialism differed from Populism and incorporated former Populists has been hotly debated, most notably by Lawrence Goodwyn and James Green. See Goodwyn, "The Cooperative Commonwealth," 8–42; James Green, "Populism, Socialism, and the Promise of Democracy," *Radical History Review* 24 (Fall 1980): 7–40. See also Donald B. Marti, "Answering the Agrarian Question: Socialists, Farmers, and Algie Martin Simons," *Agricultural History* 65 (Summer 1991): 53–69.

78. Eugene Debs, "The Socialist Party—What It Stands For," *Appeal to Reason,* January 4, 1913; "What Socialists Want to Accomplish," *Appeal to Reason,* April 27, 1912; Donald T. Critchlow, ed., *Socialism in the Heartland: The Midwestern Experience, 1900–1925* (Notre Dame, Ind.: University of Notre Dame Press, 1986), 1–11; John Graham, ed., *"Yours for the Revolution": The "Appeal to Reason", 1895–1922* (Lincoln: University of Nebraska Press, 1990), ix–x.

79. James R. Green, *Grass-Roots Socialism: Radical Movements in the Southwest 1895–1943* (Baton Rouge: Louisiana State University Press, 1978); Garin Burbank, *When Farmers Voted Red: The Gospel of Socialism in the Oklahoma Countryside, 1910–24* (Westport, Conn.: Greenwood, 1976).

80. *Bee Nugget,* July 4, 1912, June 20, 1912, May 26, 1911, January 4, 1912, January 28, 1912, February 29, 1912.

81. Ibid., April 14, 1911.

82. Minute Book, June 1, 1912, August 17, 1912, Hope Grange; *Appeal to Reason,* January 18, 1913; U.S. Census manuscripts for Eden precinct, 1900 and 1910.

83. *Bee Nugget,* January 30, 1913.

84. Ibid., January 16, 1913, April 21, 1911, April 28, 1911; *Toledo Community Story,* 89.

85. *Bee Nugget,* January 9, 1913.

86. Ibid., January 16, 1913.

87. Ibid., May 12, 1911.

88. Ibid., April 28, 1911.

89. *Appeal to Reason,* June 8, 1912, January 27, 1912; Marti, "Answering the Agrarian Question." The *Appeal,* it should be noted, was not an official mouthpiece of the Socialist party, but a strong supporter of the party; it regularly received contributions from Eugene Debs.

90. Jenny Pelto, "The Finns of Winlock, Washington," in Lillian Maki, Mrs. Bill Weible, Carl Poutala, eds., *Finns of Winlock, Lewis County Washington* (Portland: Finnish-American Historical Society of the West, 1979), 3–4; Anton Palo and Ethel Maki, "History of Winlock U.F.K.B.& S. Lodge #19," in *United Finnish Kaleva Brothers and Sisters History 1886–1979* (Astoria, Oreg.: Consolidated Printing, 1979), 17; U.S. Census manuscripts for Lewis county, 1910.

91. Pelto, "Finns of Winlock," 3–5, 12, 18–19; "History of Winlock U.F.K.B.& S. Lodge #19," 16; Membership Roll Book, 1908–90, Winlock Finnish Lodge; *Winlock News,* June 29, 1923, July 20, 1923.

92. At the time of the 1910 census, 40 percent of the population of Ainslie over age sixteen was Finnish, 26.7 percent of Veness, and 11.9 percent of Prescott. The only direct evidence we have of Socialist meetings among local Finns is in Jenny Pelto's memoir, "Finns of Winlock," 13. On Finnish immigrant radicalism, see Peter Kivisto, *Immigrant Socialism in the United States: The Case of Finns and the Left* (Cranbury, N.J.: Associated University Presses, 1984), 21–28, 89–94; Gary Marks and Matthew Burbank, "Immigrant Support for the American Socialist Party, 1912 and 1920," *Social Science History* 14 (Summer 1990): 175–202; Auvo Kostiainen, *The Forging of Finnish-American Communism, 1917–1924* (Turku: Turun Yliopisto, 1978), 19, 35–39; Carl Ross, *The Finn Factor in American Labor, Culture, and Society* (New York Mills, Minn.: Parta Printers, 1977), 56, 70–74; Al Gedicks, "The Social Origins of Radicalism among Finnish Immigrants in Midwest Mining Communities," *Review of Radical Political Economics* 8 (Fall 1976): 1–31.

93. *Bee Nugget,* May 6, 1912, August 22, 1912, September 5, 1912, September 12, 1912, September 26, 1912, October 10, 1912.

94. Marks and Burbank also stress the importance of examining rational responses to political choices when studying immigrant voting behavior, in addition to the "cultural baggage" and immigrant experience that many historians emphasize. Marks and Burbank, "Immigrant Support for the American Socialist Party," 177–78.

95. Six of the eight precincts supporting the Socialist ticket in 1912 had also given a substantial vote to the Socialists in the 1908 presidential race and the 1910

congressional campaign. Socialist support remained strong through 1916 in Veness, Ainslie, Eden, and Ethel, with between 20 percent and 34 percent of the vote. The Socialist vote fell to just under 15 percent in Cinebar and Salkum by 1916, and to under 10 percent in Prescott and Granite. The sharp decline of Socialism in Granite can be attributed to the sudden influx of nonfarmers with the building of a large mill and the construction of the town of Onalaska in 1914. No doubt some of that effect spread to neighboring Salkum and Cinebar as well.

96. Soike also found post-Civil War Norwegian settlements to have variable voting patterns. Soike, *Norwegian Americans*, 188. Others have found distinctive German and Scandinavian party voting patterns. See Marks and Burbank, "Immigrant Support for the American Socialist Party," 176.

97. For discussions of reactionary manifestations of Populism, see Nancy Mac-Lean, "The Leo Frank Case Reconsidered: Gender and Sexual Politics in the Making of Reactionary Populism," *The Journal of American History* 79 (December 1991): 917–48; C. Vann Woodward, *Tom Watson, Agrarian Rebel* (New York: Macmillan Company 1938).

6. Specialization and Cooperation: Agricultural Change in the Early Twentieth Century

1. For discussion of stable and post-frontier agricultural communities see Hal S. Baron, *Those Who Stayed Behind: Rural Society in 19th Century New England* (Cambridge: Cambridge University Press, 1984); Nancy Grey Osterud, *Bonds of Community: The Lives of Farm Women in Nineteenth-Century New York* (Ithaca: Cornell University Press, 1991); Mary Neth, "Preserving the Family Farm: Farm, Family, and Community in the Midwest, 1900–1940" (Ph.D. diss., University of Wisconsin-Madison, 1987); Jane Marie Pederson, *Between Memory and Reality: Family and Community in Rural Wisconsin, 1870–1970* (Madison: University of Wisconsin, 1992); Carol K. Coburn, *Life at Four Corners: Religion, Gender, and Education in a German-Lutheran Community, 1868–1945* (Lawrence: University Press of Kansas, 1992); Allan G. Bogue, *From Prairie to Corn Belt: Farming on the Illinois and Iowa Prairies in the Nineteenth Century* (Chicago: University of Chicago Press, 1963); Paula M. Nelson, *After the West was Won: Homesteaders and Town-Builders in Western South Dakota 1900–1917* (Iowa City: University of Iowa Press, 1986).

2. U.S. Census summary population volumes and manuscripts for Lewis County, 1900 and 1910.

3. James H. Shideler, *Farm Crisis 1919–1923* (Berkeley: University of California Press, 1957), 1–5; Gilbert C. Fite, *American Farmers: The New Minority* (Bloomington: Indiana University Press, 1981), 31.

4. Throughout the study area, property holders in 1910 who had lived in the same precinct in 1900 had assessed values averaging 70 percent more than new-comers with property. For the Silver Creek and Winlock area precincts combined, the average assessed real property value for property holders counted in both 1900 and 1910 was $960, and for people identified only in 1910 it was $564. In the entire study area, 24.4 percent of new households living on farms held mortgages, compared to 8.9 percent of households present in the same precinct since 1900, and

20.5 percent for households that had moved between study precincts. These and subsequent figures are based on combined U.S. Census manuscripts for 1900 and 1910, and Lewis county assessor records for 1900 and 1909, Washington State Archives, Olympia.

5. U.S. Department of Agriculture, "Farming the Logged-Off Uplands in Western Washington," by E. R. Johnson and E. D. Strait, Department Bulletin #1236 (Washington, D.C.: GPO, July 1924), 16, 21, 23. This study also found that settlers who had arrived after 1898 were more likely to have mortgages than those who had come before, not only because they had recently purchased their land, but also because many came from business or professional backgrounds and were used to borrowing for investments.

6. Alma Nix and John Nix, eds., *The History of Lewis County Washington* (Chehalis: Lewis County Historical Society, 1985), 141, 318; Roll Book, Hope Grange 1904–19; Jenny Pelto, "The Finns of Winlock, Washington," in Lillian Maki, Mrs. Bill Weible, Carl Poutala, eds., *Finns of Winlock* (Portland, Oreg.: Finnish-American Historical Society of the West, April 1979), 4. In Ainslie, eighteen heads of households living on farms worked on the farm and twenty-three worked off in 1910. In Veness, thirty-two heads of farm households worked on the farm and thirty worked off.

7. From U.S. Census, *Agricultural Statistics*, 1890, 1900, 1910, 1920.

8. Richard White, *Land Use, Environment, and Social Change: The Shaping of Island County, Washington* (Seattle: University of Washington Press, 1980), chap. 3.

9. USDA Bulletin #1236, 2; U.S. Department of Agriculture, "The Utilization of Logged-Off Land for Pasture in Western Oregon and Western Washington," by Byron Hunter and Harry Thompson, USDA Farmers' Bulletin #462 (Washington, D.C.: GPO, 1911). Richard White's *Land Use* is the best general book on the relationship between western Washington forests and European-style agriculture. See also Campbell Garrett Murphy, "Farmers on Cut-Over Timber Lands in Western Washington State: A Study of the Related Programs of Governmental Agencies Concerned with Their Rehabilitation with Special Reference to the Farm Security Administration" (master's thesis, University of Washington, 1943).

10. USDA Bulletin #1236, 5, 16.

11. For discussion of the Country Life movement, see Roy V. Scott, preface to *The Reluctant Farmer: The Rise of Agricultural Extension to 1914* (Urbana: University of Illinois, 1970); David B. Danbom, *The Resisted Revolution: Urban America and the Industrialization of Agriculture, 1900–1930* (Ames: Iowa State University, 1979), 10–13; William L. Bowers, *The Country Life Movement in America 1900–1920* (Port Washington, N.Y.: Kennikat Press, 1974); and Neth, "Preserving the Family Farm," 126–72. For discussions of farming adaptations in other contexts see Eric E. Lampard, *The Rise of the Dairy Industry in Wisconsin: A Study in Agricultural Change, 1820–1920* (Madison: State Historical Association of Wisconsin, 1963); Mark Friedberger, *Farm Families and Change in Twentieth-Century America* (Lexington: University of Kentucky Press, 1988); J. Sanford Rikoon, *Threshing in the Midwest, 1820–1940: A Study of Traditional Culture and Technological Change* (Bloomington: Indiana University Press, 1988); Pete Daniel, *Breaking the Land: The Transformation of Cotton, Tobacco, and Rice Cultures Since 1880* (Urbana: University of Illinois Press, 1985).

12. *Chehalis Bee*, November 6, 1891, November 20, 1891.

13. Quotation in *People's Advocate*, August 2, 1895; see also September 21, 1894, February 22, 1895.

14. See, for example, *Bee Nugget*, December 28, 1900, April 5, 1901, August 12, 1910, July 6, 1911. See also Minnie Lingreen and Priscilla Tiller, *Hop Cultivation in Lewis County, Washington, 1888–1940: A Study in Land Use Determinants* (Centralia, Wash.: 1981), 33, 39.

15. *Chehalis Bee*, January 22, 1892, June 10, 1892; *Bee Nugget*, February 1, 1901.

16. *Bee Nugget*, November 4, 1910, January 13, 1911, May 26, 1911.

17. Ibid., August 2, 1918.

18. Ibid., April 30, 1920, August 13, 1920.

19. Ibid., May 27, 1921, August 19, 1921. For discussion of the farm crisis following World War I, see Shideler, *Farm Crisis*, 46–52; Fite, *American Farmers*, 34–37; Richard Franklin Bensel, *Sectionalism and American Political Development 1880–1980* (Madison: University of Wisconsin Press, 1984), chap. 4.

20. *Winlock News*, January 12, 1923, April 6, 1923, June 8, 1923, June 22, 1923, July 13, 1923.

21. Ibid., February 2, 1923, February 16, 1923.

22. *Bee Nugget*, January 23, 1912, August 1, 1913.

23. *People's Advocate*, December 23, 1892, July 14, 1893, August 18, 1893, April 10, 1896, June 12, 1896, May 7, 1897, June 4, 1897.

24. Ibid., February 3, 1899, March 3, 1899; E. F. Dummeier, "Co-operation in Marketing Washington Farm Products," Agricultural Experiment Station, Bulletin #194 (Pullman, Wash.: State College of Washington, 1925), 15. Even after 1921 some cooperatives continued to organize under the private corporation law because of the restrictions imposed by the Cooperative Marketing Act.

25. Quotation in *Bee Nugget*, November 30, 1900; see also April 26, 1901, May 3, 1901; *People's Advocate*, April 14, 1899.

26. *Bee Nugget*, March 7, 1913.

27. Ibid., March 7, 1913, May 30, 1913.

28. Ibid., March 28, 1912, April 25, 1912, May 2, 1912, May 9, 1912, September 26, 1912, June 20, 1913.

29. *Bee Nugget*, February 21, 1919, June 18, 1920, February 3, 1922; Agricultural Experiment Bulletin #194, 53–54.

30. *Bee Nugget*, January 17, 1919; Articles of Incorporation and Trustees Meeting, May 10, 1919, Lewis-Pacific Dairymen's Association (LPDA) Papers, Lewis County Historical Society, Chehalis, Wash.

31. Stockholders' Meeting, January 23, 1922, LPDA Papers; Agricultural Experiment Station Bulletin #194, 58–59.

32. *Bee Nugget*, July 9, 1920, April 22, 1921; Manager's Report, January 28, 1924, LPDA Papers.

33. Bylaws 1919, LPDA Papers.

34. *Bee Nugget*, October 1, 1920, October 8, 1920, November 5, 1920.

35. Ibid., February 18, 1921, September 2, 1921, September 23, 1921.

36. Ibid., April 1, 1921, January 13, 1922; Stockholders' Meeting, January 23, 1922, LPDA Papers.

37. *Bee Nugget*, December 23, 1921, January 13, 1922.

38. Ibid., January 6, 1922, February 3, 1922.

39. Agricultural Experiment Station Bulletin #194, 55.

40. *Bee Nugget*, March 18, 1921, June 16, 1922; Agricultural Experiment Station Bulletin #194, 54.

41. *Bee Nugget*, February 24, 1922, February 2, 1923; C. C. Wall, "A History of Winlock," 1952, 58, Winlock High School; *The Toledo Community Story, 1853–1953*, 157–58, Seattle Public Library.

42. *Bee Nugget*, March 29, 1918, February 6, 1920.

43. Ibid., March 11, 1921.

44. *Bee Nugget*, March 29, 1918. Probably the majority of LPDA members belonged to the Grange. A stockholders list from 1920 gave no indication where in the two-county area people lived, but eighty-two of those names were located in the 1910 census of the study-area precincts. Forty-five percent of that number appeared on Grange rolls that could be located. That percentage is a minimum number— the actual percentage was probably higher given the incomplete nature of the data and the ten-year lag with the census data.

45. Elizabeth Faue makes a similar distinction between workplace-centered, bureaucratic labor unions and community-based organizing in *Community of Suffering and Struggle: Women, Men, and the Labor Movement in Minneapolis, 1915–1945* (Chapel Hill: University of North Carolina Press, 1991), 1–14. In Faue's analysis, mobilization that included the entire working-class community was potentially far more radical than unions that focused on only the economic roles of select male workers.

46. *Bee Nugget*, November 30, 1900, December 13, 1900, April 21, 1911, August 10, 1911, November 21, 1913, December 20, 1918; *People's Advocate*, December 14, 1900.

47. George Severance and Ernest R. Johnson, "The Cost of Producing Milk, and Dairy Farm Organization in Western Washington," Agricultural Experiment Station, Bulletin #173, (Pullman, Wash.: State College of Washington, 1922), 36; Annual Report of Lewis County Agent A. B. Nystrom, Chehalis, Wash., 1918, Extension Service Annual Reports, National Archives, Pacific Northwest Branch.

48. Minute Book, April 20, 1918, May 4, 1918, St. Urban Grange, Winlock.

49. Annual Report of Lewis County Agent A. B. Nystrom 1918; Annual Report of Assistant Dairy Specialist F. W. Kehrli, 1922; Annual Report of Lewis County Agent Francis D. Yeager, 1924, Extension Service Annual Reports, National Archives.

50. *Bee Nugget*, June 21, 1918, April 2, 1920, April 9, 1920, January 20, 1922.

51. Minute Book, February 15, 1919, May 17, 1919, St. Urban Grange.

52. Agricultural Experiment Bulletin #173, 5, 10, 14, 31, 33 (quotation).

53. Nix and Nix, eds., *History of Lewis County*, 142–43; *Bee Nugget*, February 4, 1921; Ethel Maki, "Winlock, Washington," in Lillian Maki, Mrs. Bill Weible, and Carl Poutala, eds., *Finns of Winlock, Lewis County Washington* (Portland, Oreg.: Finnish-American Historical Society of the West, 1979), 18. Carolyn E. Sachs, *The Invisible Farmer: Women in Agricultural Production* (Totowa, N.J.: Rowman & Allanheld, 1983), 40–41, describes the same general pattern throughout the United States in chicken raising—from household food raised by women to large-scale commercial enterprises relying heavily on technology.

54. *Bee Nugget*, November 30, 1900, February 8, 1901, February 15, 1901.

55. Wall, "History of Winlock," 53; *Bee Nugget*, December 8, 1922.

56. *Bee Nugget*, February 11, 1921, March 16, 1923, April 20, 1923, June 29, 1923.

57. Ibid., March 12, 1920, June 25, 1920, July 23, 1920.

58. Agricultural Experiment Bulletin #194, 32–33.

59. *Lewis County Advocate,* March 17, 1922.

60. Agricultural Experiment Bulletin #194, 39, 41; *Winlock News,* January 19, 1923, April 13, 1923.

61. Quotation in *Winlock News,* April 13, 1923; see also July 6, 1923.

62. Ibid., June 29, 1923, July 6, 1923.

63. Wall, "History of Winlock," 53.

64. Various perspectives on the Country Life Movement and the rise of the Extension Service are provided in Bowers, *Country Life Movement;* Scott, *The Reluctant Farmer;* Neth, "Preserving the Family Farm"; Danbom, *The Resisted Revolution.*

65. Extension Department Annual Report, W. S. Thurnber, December 1915, 2, Extension Service Annual Reports, National Archives; Scott, *The Reluctant Farmer,* chap. 3.

66. *Bee Nugget,* October 10, 1913, October 31, 1913, November 14, 1913, November 21, 1913, January 9, 1914.

67. Annual Report of County Agent Leader R. B. Coglon, State of Washington, 1917, 15–16, Extension Service Annual Reports, National Archives.

68. Annual Report of Lewis County Agricultural Agent A. B. Nystrom, December 1917, Extension Service Annual Reports, National Archives.

69. Annual Report of Lewis County Agent A. B. Nystrom, 1918, 27; Annual Report of Mary E. Sutherland, Clothing and Food Specialist, January 1920; Annual Report of Maud Wilson, State Home Demonstration Leader, 1921, Extension Service Annual Reports, National Archives. For discussion of gender ideology in the Extension Service see Katherine Jellison, *Entitled to Power: Farm Women and Technology, 1913–63* (Chapel Hill: University of North Carolina Press, 1993); Neth, "Preserving the Family Farm," 201; Deborah Fink, *Open Country Iowa: Rural Women, Tradition, and Change* (Albany: SUNY Press, 1986), 99; Sachs, *The Invisible Farmer,* 60; Joan M. Jensen, *With These Hands: Women Working on the Land* (Old Westbury, N.Y.: The Feminist Press, 1981), 153; Dorothy Schwieder, "Education and Change in the Lives of Iowa Farm Women, 1900–1940," *Agricultural History* 60 (Spring 1986): 200–215; Kathleen R. Babbitt, "The Productive Farm Woman and the Extension Home Economist in New York State, 1920–1940," *Agricultural History* 67 (Spring 1993): 83–101.

70. Samuel R. Berger, *Dollar Harvest: The Story of the Farm Bureau* (Lexington, Mass.: D.C. Heath, 1971) gives a highly critical view of the Farm Bureau. Robert P. Howard, *James R. Howard and the Farm Bureau* (Ames: Iowa State University, 1983) is a laudatory history written by the founder's son. Both make clear the Farm Bureau's antiradical stance. See also John Mark Hansen, *Gaining Access: Congress and the Farm Lobby, 1919–1981* (Chicago: University of Chicago Press, 1991), 26–77.

71. Lewis County Annual Report of A. B. Nystrom, 1919, 4, Extension Service Annual Reports, National Archives.

72. Annual Report of A. T. Flagg, Lewis County, 1920, 4; Annual Report of Lewis County Agent R. G. Fowler, 1922, 3, Extension Service Annual Reports, National Archives.

73. "Lewis County," list of subordinate granges, Washington State Grange, Olympia, Washington, n.d.

74. Narrative and Statistical Report of County Agent Leader, R. B. Coglon, State

of Washington, December 1, 1917 to December 1, 1918, 4, Extension Service Annual Reports, National Archives.

75. Report of County Agent Leader R. B. Coglon, State of Washington, 1916, 3, Extension Service Annual Reports, National Archives.

76. Annual Report of Lewis County Agent R. G. Fowler, October 15, 1922, 5, Extension Service Annual Report, National Archives.

77. Nix and Nix, eds., *History of Lewis County*, 142–43. A USDA survey of over 10,000 farm women in 1920 found that women cared for eighty-one percent of all poultry flocks in the country, according to the *Agricultural Grange News*, July 5, 1920. Both Carolyn Sachs and Deborah Fink in their studies of farm women's work describe such an evolution, the same that occurred in the Erving family near Winlock in the 1910s. Sachs, *The Invisible Farmer*, 40–41; Fink, *Open Country, Iowa*, chap. 6. In a study of the rise of the dairy industry in Wisconsin, Eric E. Lampard states that dairying had always been considered women's work until certain processes were industrialized and it became a major money-maker. Then men took over the business. Osterud, on the other hand, found that the division of labor in dairy work varied significantly by family in late-nineteenth-century New York, and that in many families husband and wife shared the work, even after it gained in market value. She suggests that this sharing of income-producing dairy work led men to have a greater respect for women's farm labor. Lampard, *Rise of the Dairy Industry*, pt. 1; Osterud, *Bonds of Community*, 150–56.

78. "Alpha Grange History," undated, 3, Alpha Grange. St. Urban debated the same question in 1921. Minute Book, September 23, 1921, St. Urban Grange.

79. Minute Book, June 3, 1922, Cougar Flat Grange; *Bee Nugget*, January 20, 1922, February 3, 1922; *Winlock News*, April 28, 1922.

80. USDA Bulletin #1236, 10–11.

81. See, for example *People's Advocate*, February 25, 1898, February 17, 1899; *Bee Nugget*, October 5, 1911, March 21, 1913.

82. Deborah Fink, *Agrarian Women: Wives and Mothers in Rural Nebraska, 1880–1940* (Chapel Hill: University of North Carolina Press, 1992), 4–5, 189–96, emphasizes the grueling labor, poverty, and isolation of Nebraska women, and concludes that the hardships of farm life far outweighed any potentially "liberating" effects of contributing to the family economy. Katherine Jellison, "Women and Technology on the Great Plains, 1910–40," *Great Plains Quarterly* (Summer 1988): 145–57, discusses conflicts between husbands and wives over spending discretionary money and gaining leisure, and at least implies that women felt their husbands undervalued their contributions. Melody Graulich, "Violence Against Women: Power Dynamics in Literature of the Western Family," in Susan Armitage and Elizabeth Jameson, eds., *The Women's West* (Norman: University of Oklahoma Press, 1987), 111–25, perceives a pervasive ethic of violence against women in literature from turn-of-the-century Plains states.

7. A Community in Conflict: The End of Tolerance

1. U.S. Census manuscripts, Little Falls precinct, 1910, households #49, 56, 120, and 124; Winlock precinct, households #175, 183, 217, and 245. The statistics that

follow are based on analysis of the 1910 census manuscripts for all the study precincts.

2. *Bee Nugget,* June 10, 1910.

3. Ibid., July 6, 1911, July 27, 1911.

4. Ibid., February 4, 1921.

5. Ibid., July 27, 1911.

6. *Winlock News,* January 12, 1923, July 13, 1923, June 15, 1923, July 13, 1922, February 2, 1923, May 18, 1923.

7. *Bee Nugget,* February 25, 1910, April 15,1910; *Winlock News,* April 6, 1923, April 27, 1923.

8. C. C. Wall, "A History of Winlock, Washington," 1952, 51, Winlock High School; *Winlock News,* July 18, 1913, January 13, 1922, June 16, 1922, December 15, 1922, March 23, 1923, April 13, 1923, May 4, 1923.

9. *Winlock News,* February 2, 1923.

10. R. L Polk & Co., Directories, *Lewis and Pacific County Directory 1910–1911, 1915–1916; Lewis County 1922–1923, 1925–1926* (Seattle: R. L. Polk & Co.).

11. I have linked names from three different types of records: the manuscripts from the 1910 federal census; Lewis county tax assessment records for 1909 (the 1910 records are not in the state archives, but taxes assessed in 1909 were paid in 1910); and organizational membership lists. Altogether, 87 Masons, 120 Eastern Stars, 14 Oddfellows, 71 Rebekahs, and 38 Woodmen who were active between 1900 and 1925 were linked to their 1910 census record. The names of Mason, Eastern Star, and Rebekah members were drawn largely from membership records of their respective organizations, which included everyone who joined, while the names of Oddfellows and Woodmen came only from newspaper accounts which primarily included lodge leaders.

12. Others who have found a similar mix of middle-class values and exclusion of those deemed "undesirable," along with a blurring of class lines in fraternal lodges, include Mary Ann Clawson, *Constructing Brotherhood: Class, and Fraternalism* (Princeton: Princeton University Press, 1989), 88–110; Elizabeth Ann Jameson, "High-Grade and Fissures: A Working-Class History of the Cripple Creek, Colorado Gold Mining District, 1890–1905" (Ph.D. diss., University of Michigan, 1987), 208–13. On the other hand, an analysis of lodges as more exclusively middle class is given in Mark C. Carnes, *Secret Ritual and Manhood in Victorian America* (New Haven: Yale University Press, 1989), 26, 31–32.

13. These descriptions are drawn from files I maintained for all individuals whose names were listed on organizational membership roles or in newspaper accounts associated with an organization. I also added personal information to these files from the 1910 manuscript census, county assessor records, and Polk directories. This information remains incomplete, both with respect to each individual and to the organizations.

14. The statistics presented here are based on those known organization members whose names could be located in the 1910 census manuscripts. Cross-membership figures need to be interpreted cautiously because complete membership lists were not available for many of the organizations, and differences in spelling and women's name changes make tracing people difficult. The percentage of members from one organization whose names were also found in other groups gives us an

indication of the relative level of joint participation only and should not be taken to represent the actual figures.

15. Limited church records were available from Cowlitz Prairie Baptist Church, Winlock (earlier Winlock First Baptist Church); St. Paul's Lutheran Church, Winlock; Toledo Presbyterian Church; and the Methodist Episcopal Churches of Winlock, Toledo, Salkum, and Ferry, records for which are kept at Winlock Methodist Episcopal Church. Other information on people involved in church life was drawn from newspaper accounts.

16. These figures are based on those known organization members located in the 1910 census manuscripts.

17. Wilson H. Grabell, Clyde V. Kiser, and Pascal K. Whelpton, "A Long View," in Michael Gordon, ed., *The American Family in Social-Historical Perspective* (New York: St. Martin's Press, 1973), 387; Daniel Scott Smith, "Family Limitation, Sexual Control, and Domestic Feminism," in Nancy F. Cott and Elizabeth H. Pleck, eds., *A Heritage of Her Own: Toward a New Social History of American Women* (New York: Touchstone, 1979), 226. Many studies of fertility decline have focused on land availability, which was relatively high in Lewis county. For an overview and assessment, see Maris A. Vinovskis, "Historical Perspectives on Rural Development and Human Fertility in Nineteenth-Century America," in Wayne A. Schutjer and C. Shannon Stokes, eds., *Rural Development and Human Fertility* (New York: Macmillan, 1984), 77–96.

18. For example, see Minute Book, April 12, 1921, December 11, 1923, Toledo Eastern Star.

19. *Winlock News*, March 23, 1923, May 4, 1923.

20. Camp life is described in a number of sources, including Robert L. Tyler, *Rebels of the Woods: The I.W.W. in the Pacific Northwest* (Eugene: University of Oregon Press, 1967); Harold M. Hyman, *Soldiers and Spruce: Origins of the Loyal Legion of Loggers and Lumbermen* (Los Angeles: University of California Press, 1963); John McClelland, Jr., *Wobbly War: The Centralia Story* (Tacoma: Washington State Historical Society, 1987); Harvey O'Connor, *Revolution in Seattle: A Memoir* (Seattle: Left Bank Books, 1981 [1964]), 59–64; Andrew Mason Prouty, *More Deadly Than War: Pacific Coast Logging 1827–1981* (New York: Garland, 1985).

21. *Bee Nugget*, April 26, 1901, May 3, 1901, May 10, 1901, October 31, 1912, February 4, 1921, December 22, 1922; Carlos A. Schwantes, *Radical Heritage: Labor, Socialism, and Reform in Washington and British Columbia, 1885–1917* (Vancouver, B.C.: Douglas & McIntyre, 1979), 153. For accounts of other mill towns, see also Norman H. Clark, *Mill Town: A Social History of Everett from Its Beginnings to the Massacre* (Seattle: University of Washington Press, 1970); William G. Robbins, *Hard Times in Paradise: Coos Bay, Oregon, 1850–1986* (Seattle: University of Washington Press, 1988); Jeremy W. Kilar, *Michigan's Lumbertowns: Lumbermen and Laborers in Saginaw, Bay City, and Muskegon, 1870–1905* (Detroit: Wayne State University Press, 1990).

22. *Bee Nugget*, December 21, 1900, March 22, 1901, April 5, 1901, October 31, 1912. For information on market swings and efforts to organize, see also Robbins, *Hard Times*, 48–50; Schwantes, *Radical Heritage*, 153–54.

23. McClelland, *Wobbly War*, 29–30; Tyler, *Rebels of the Woods*, 92; Hyman, *Soldiers and Spruce*, 48–65.

24. Clark, *Mill Town*, 186–214; McClelland, *Wobbly War*, 11–14.

25. The attack on civil liberties during and immediately following World War I

is well documented. See Melvin Dubofsky, *We Shall Be All: A History of the Industrial Workers of the World* (New York: Quadrangle, 1969), 438–45; Tyler, *Rebels of the Woods*, 116–54; McClelland, *Wobbly War*, 32. In Washington, an antisyndicalism law passed by the legislature in 1917 was vetoed by Governor Lister but finally approved in 1919.

26. *Toledo Messenger*, July 19, 1917.

27. *Bee Nugget*, February 22, 1918.

28. Ibid., January 11, 1912, December 5, 1913; McClelland, *Wobbly War*, 1.

29. *Bee Nugget*, April 26, 1918; *Toledo Messenger*, April 18, 1918; Tom Copeland, *The Centralia Tragedy of 1919: Elmer Smith and the Wobblies* (Seattle: University of Washington Press, 1993), 36–38. Although local newspaper reports clearly place this parade and attack in April, some historians have mistakenly said it occurred in May, including McClelland, *Wobbly War*, 51–53; Kerry Irish, "Eternal Vengeance: A History of the Centralia Massacre" (master's thesis, University of Washington, 1989), 17.

30. *Toledo Messenger*, July 12, 1917; *Bee Nugget*, May 10, 1918, June 7, 1918.

31. Accounts of the Centralia Massacre vary widely. The most recently published works still differ in some details, but are reasonably balanced: Copeland, *Centralia Tragedy*; McClelland, *Wobbly War*. For a totally different view, see Raymond Moley, Jr., *The American Legion Story* (New York: Duell, Sloan, & Pearce, 1966), 98–100.

32. *Bee Nugget*, November 14, 1919.

33. Copeland, *Centralia Tragedy*, 55–59; McClelland, *Wobbly War*, 95–105.

34. *Winlock News*, April 27, 1923; *Bee Nugget*, February 6, 1920, March 12, 1920, June 3, 1921, June 10, 1921.

35. *Winlock News*, February 24, 1922; Copeland, *Centralia Tragedy*, 178–180.

36. *Toledo Messenger*, August 9, 1917; see also all the issues for July and August 1917.

37. The Washington Grange was not the only other victim. In addition to nearly destroying the IWW, the postwar Red Scare also came close to stamping out the Socialist party. See David H. Bennett, *The Party of Fear: From Nativist Movements to the New Right in American History* (Chapel Hill: University of North Carolina Press, 1988) 183–97; John Higham, *Strangers in the Land: Patterns of American Nativism 1860–1925* 2d ed. (New York: Atheneum, 1974), 222–33; James R. Green, *Grass-Roots Socialism: Radical Movements in the Southwest 1895–1943* (Baton Rouge: Louisiana State University Press, 1978), 382.

38. For histories of the NPL, see Charles Edward Russell, *The Story of the Non-Partisan League: A Chapter in American Evolution* (New York: Harper, 1920); Kathleen Diane Moum, "Harvest of Discontent: The Social Origins of the Nonpartisan League, 1880–1922" (Ph.D. diss., University of California, Irvine, 1986); Richard M. Valelly, *Radicalism in the States: The Minnesota Farmer-Labor Party and the American Political Economy* (Chicago: University of Chicago Press, 1989), 17–32.

39. Moum, "Harvest," 193; Russell, *Story of the NPL*, 231.

40. See Harriet Ann Crawford, *The Washington State Grange, 1889–1924: A Romance of Democracy* (Portland, Oreg.: Binfords & Mort, 1940), 233.

41. *Toledo Messenger*, May 2, 1918.

42. *Bee Nugget*, May 3, 1918, May 10, 1918; *Toledo Messenger*, May 2, 1918. The KKK resurgence during this period is discussed in Bennett, *The Party of Fear*, 208–36; Kathleen M. Blee, *Women of the Klan: Racism and Gender in the 1920s* (Berkeley:

University of California Press, 1991), 154–73; Shawn Lay, ed., *The Invisible Empire in the West: Toward a New Historical Appraisal of the Ku Klux Klan of the 1920s* (Urbana: University of Illinois Press, 1992); Jeff LaLande, "Beneath the Hooded Robe: Newspapermen, Local Politics, and Ku Klux Klan in Jackson County, Oregon, 1921–1923," *Pacific Northwest Quarterly* 83 (April 1992): 42–52.

43. *Bee Nugget,* May 10, 1918.

44. Ibid.

45. Ibid., May 24, 1918, May 31, 1918, June 21, 1918.

46. Ibid., June 14, 1918, June 21, 1918, June 28, 1918; McClelland, *Wobbly War,* 62.

47. *Proceedings of the Washington State Grange. 1918,* 18–19; letter from Alice Yarnell, Pierce County Pomona Grange delegate, published in *Tacoma Labor Advocate,* June 21, 1918.

48. *Proceedings of the Grange, 1918,* 19.

49. *Proceedings of the Grange, 1920,* 19; Carlos A. Schwantes, "Farmer-Labor Insurgency in Washington State: William Bouck, the Grange, and the Western Progressive Farmers," *Pacific Northwest Quarterly* 76 (1985): 2–11.

50. Schwantes, "Farmer-Labor Insurgency," 4–7; Crawford, *Washington State Grange,* 235, 271; *Union Record,* March 6, 1919, February 10, 1920, October 9, 1920.

51. Minute Book, July 6, 1918, St. Urban Grange; Minute Book, June 27, 1918, January 27, 1919, Alpha Grange; Minute Book, December 6, 1918, October 16, 1920, Silver Creek Grange; Minute Book, October 13, 1921, October 27, 1921, Cowlitz Prairie Grange.

52. *Bee Nugget,* September 16, 1921 (quotation), August 12, 1921.

53. Ibid., August 12, 1921, (long quotation), June 17, 1921 ("past understanding" statement), November 25, 1921.

54. In 1919 there were 1,179 Grange members in Lewis county and 15,246 in the state, and in 1921 1,678 in Lewis county and 20,736 in the state. *Proceedings of the Grange, 1919,* 79, *1922,* 57.

55. *Bee Nugget,* April 14, 1922; Minute Book, May 17, 1919, December 4, 1919, Ethel Grange; Minute Book, June 5, 1920, Silver Creek Grange; Minute Book, December 2, 1918, September 20, 1919, May 3, 1919, August 2, 1919, August 16, 1919, St. Urban Grange; Minute Book, August 19, 1922, Cougar Flat Grange; Minute Book, April 12, 1919, June 7, 1919, Hope Grange.

56. *Bee Nugget,* April 9, 1920 (quotation); Alpha Grange passed a nearly identical resolution. Minutes Book, March 27, 1920, March 13, 1920, Alpha Grange. See also Minute Book, December 4, 1919, Ethel Grange; Minute Book, February 8, 1919, Hope Grange.

57. *Toledo Messenger,* May 2, 1918; *Bee Nugget,* February 20, 1920, March 12, 1920, March 26, 1920, May 21, 1920, May 28, 1920.

58. *Bee Nugget,* July 2, 1920.

59. Hamilton Cravens, "A History of the Washington Farmer-Labor Party, 1918–1924" (master's thesis, University of Washington, 1962), 82–118, 130; Robert Donald Saltvig, "The Progressive Movement in Washington" (Ph.D. diss., University of Washington, 1966), 468; Schwantes, "Farmer-Labor Insurgency," 5–6.

60. Secretary of State, General Election Returns, Lewis County, 1920, Washington State Archives, Olympia; Cravens, "Farmer-Labor Party," 137.

61. *Bee Nugget*, September 3, 1920, October 15, 1920, October 22, 1920.

62. *Proceedings of the Grange, 1921, 1923*; Cravens, "Washington Farmer-Labor Party," 129–45, 163–84; Saltvig, "Progressivism," 468–79.

63. See Cravens, "Washington Farmer-Labor Party," 139; Saltvig, "Progressivism," 479; Schwantes, "Farmer-Labor Insurgency," 11.

64. Secretary of State, General Election Returns, 1922, Washington State Archives, Olympia.

65. General Election Returns, 1922. The other three Granges that closed during this period were Cinebar, Napavine, and Producers near Winlock. "Lewis County," list of subordinate granges, Washington State Grange, Olympia, n.d.

66. *Lewis County Advocate*, October 13, 1922, November 3, 1922; Jonathan Dembo, *Unions and Politics in Washington State, 1885–1935* (New York: Garland, 1983), 298–300.

67. General Election Returns, 1924; Schwantes, "Farmer-Labor Insurgency," 10.

68. Schwantes, "Farmer-Labor Insurgency," 9; *Western Progressive Farmer*, July 20, 1923.

69. *Western Progressive Farmer*, June 20, 1923.

70. Ibid., February 20, 1923.

71. Ibid., April 20, 1923 (quotation), February 20, 1923.

72. Ibid., June 20, 1923.

73. Ibid., May 20, 1923.

74. Ibid., May 20, 1923, February 20, 1923, March 20, 1923.

75. *Proceedings of the Grange, 1924*, 167.

76. *Western Progressive Farmer*, January 20, 1923, March 20, 1923, April 20, 1923 (quotation).

77. *Proceedings of the Grange, 1922*, 17–18, 72–76.

78. *Proceedings of the Grange, 1923*, 30–33; Crawford, *Washington State Grange*, 291–97.

79. *Western Progressive Farmer*, May 20, 1923.

80. See membership totals in the *Proceedings of the Washington Grange 1920–26*; Minute Book, November 4, 1922, Hope Grange (attendance averaged in the mid- to high 20s in 1921, but remained in the low teens through 1922); Membership Roll Book, Alpha Grange.

81. *Western Progressive Farmer*, September 15, 1924; Schwantes, "Farmer-Labor Insurgency," 10.

82. Minute Book, August 19, 1922, October 21, 1922, Cougar Flat Grange; Minute Book, January 5, 1924, March 11, 1924, Alpha Grange; Minute Book, February 14, 1924; Cowlitz Prairie Grange.

83. Crawford, *Washington State Grange*, 305; Schwantes, "Farmer-Labor Insurgency," 9.

84. For examples of good-road activities, see *Bee Nugget*, January 25, 1918, December 6, 1918, February 21, 1919, February 6, 1920, February 25, 1921; Minute Book, August 19, 1922, Cougar Flat Grange; Minute Book, February 16, 1918, March 15, 1919, April 3, 1920, St. Urban Grange.

85. *Winlock News*, April 13, 1923, April 27, 1923.

86. Ibid., September 8, 1922; R. L. Polk & Co., *Lewis County Directory, 1922–23* (Seattle: R. L. Polk & Co.).

87. *Winlock News*, February 2, 1923.
88. Ibid., February 9, 1923, February 16, 1923, March 9, 1923.
89. Ibid., February 2, 1923, February 16, 1923.

8. Conclusion

1. Richard M. Valelly, *Radicalism in the States: The Minnesota Farmer-Labor Party and the American Political Economy* (Chicago: University of Chicago Press, 1989); Charles Edward Russell, *The Story of the Non-Partisan League: A Chapter in American Evolution* (New York: Harper, 1920); Millard L. Gieske, *Minnesota Farmer-Laborism: The Third-Party Alternative* (Minneapolis: University of Minnesota Press, 1979); Kathleen Diane Moum, "Harvest of Discontent: The Social Origins of the Nonpartisan League, 1880–1922" (Ph.D. diss., University of California, Irvine, 1986); Richard Franklin Bensel, *Sectionalism and American Political Development 1880–1980* (Madison: University of Wisconsin Press, 1984).

2. Michael Schwartz, *Radical Protest and Social Structure: The Southern Farmers' Alliance and Cotton Tenancy, 1880–1890* (New York: Academic Press, 1976), 11, 13, 103, 110. C. Vann Woodward, *Origins of the New South, 1877–1913* (Baton Rouge: Louisiana State University Press, 1951).

3. For elaboration of these concepts, see Valelly, *Radicalism in the States*; Schwartz, *Radical Protest*; Richard Hogan, *Class and Community in Frontier Colorado* (Lawrence: University of Kansas Press, 1990); John Walton, *Western Times and Water Wars: State, Culture, and Rebellion in California* (Berkeley: University of California Press, 1992).

4. Schwartz, *Radical Protest*, 110.

5. Scott G. McNall, *The Road to Rebellion: Class Formation and Kansas Populism, 1865–1900* (Chicago: University of Chicago Press, 1988), 234. A more central role for women in the Kansas Alliance is suggested by Michael L. Goldberg, "Non-Partisan and All-Partisan: Rethinking Woman Suffrage and Party Politics in Gilded Age Kansas," *Western Historical Quarterly* 25 (Spring 1994): 21–44.

6. Angela Y. Davis, *Women, Race, and Class* (New York: Random House, 1981), 121; Jacquelyn Dowd Hall, *Revolt against Chivalry: Jesse Daniel Ames and the Campaign against Lynching* (New York: Columbia University Press, 1979).

7. Nancy Grey Osterud, *Bonds of Community: Lives of Farm Women in Nineteenth Century New York* (Ithaca: Cornell University Press, 1991); Mary Neth, "Preserving the Family Farm: Farm Families and Communities in the Midwest, 1900–1940" (Ph.D. diss., University of Wisconsin-Madison, 1987); Deborah Fink, *Open Country, Iowa: Rural Women, Tradition, and Change* (Albany: SUNY Press, 1986).

8. *Winlock News*, September 8, 1922, February 2, 1923, February 9, 1923, February 16, 1923.

Index

African-Americans, 40–44
Ainslie, 20, 25–26, 110, 127, 156, 223 n. 6; voting in, 79–80, 105, 119–20, 179, 222 n. 95
Alexander, John, 56
Alpha Prairie, 18, 21, 26, 29–31, 110, 125; voting in, 79–80, 105, 119–20, 179; Farmers' Alliance of, 54, 56; Grange of, 85, 88, 91–92, 94, 100, 102, 150, 175–77, 181, 185–86, 215 n. 9
American Legion, 93, 174, 180; and Centralia Massacre, 168
American Protective Association, 44, 75–78, 213 n. 32
Annonen, John, 142
Anthony, Susan B., 96
Antrim, Hattie and Peter, 101–2, 114
Appeal to Reason, 113–17, 221 n. 89

Baldwin, Mollie, 161
Baptists, 32–36, 44–45
Betty family, 33
Bezemer, Klaus, 29, 31, 36, 207 n. 55
Bouck, William, 94, 109, 172–76; and Western Progressive Farmers, 181–84
Bridges, Arthur, 31
Burbank, Garin, 114
Burnt Ridge, 32, 64
Bush, Judd C., 115–16, 136–38

Carnes, Mark C., 38–39, 228 n. 12
Carpenter, Clara and Ed, 24–25

Catholics, 18, 32–36; and anti-Catholicism, 44, 75–78, 208 n. 70
census taking, 24, 40, 204–5 n. 21, 208 n. 61
Centralia, 18, 76, 80, 111, 156; and IWW, 167–69, 230 n. 29
charitable activities, 37, 49, 92, 157
Chehalis, 18, 80, 111, 132–33, 156; and businessmen's organizations, 44, 80–83, 130, 137, 146; and IWW, 167–68; and women's organizations, 49
Chehalis *Bee Nugget*, 13–14, 76–77, 115
Chinese, and anti-Chinese movements, 43, 213 n. 33
Christian Church (Disciples of Christ), 32, 45
Christian Woman's Temperance Union, 30, 45
churches, 18–20, 32–36, 38, 41–45, 106, 161–64, 206–7 n. 45, 218–19 n. 53, 229 n. 15
Cinebar, 17, 26, 125; Farmers' Alliance of, 72–73; Grange of, 185; voting in, 79–80, 104–5, 119–20, 179, 222 n. 95
civic clubs, 49, 157–58, 164, 183, 187
class divisions, 10, 14, 28, 37–40, 46–47, 153–60, 163–64, 207–8 n. 60
Clawson, Mary Ann, 38–39, 228 n. 12
community base of politics, 7–13, 139, 151–52, 190–97, 215 n. 3, 225 n. 45; and Farmers' Alliance, 48–52, 58–59, 64–65; and Grange, 85–86, 94, 189; and Socialist party, 117–18